5/05

D0853419

3000 800056 40743
St. Louis Community College

Meramec Library
St. Louis Community College
11333 Big Bend Blvd.
Kirkwood, MO 63122-5799
314-984-7797

Magick, Mayhem, and Mavericks

Cathy Cobb

Magick,
Mayhem,
and
Mavericks

THE SPIRITED HISTORY OF PHYSICAL CHEMISTRY

St. Louis Community College
at Meramec
LIBRARY

 Prometheus Books

59 John Glenn Drive
Amherst, New York 14228-2197

Published 2002 by Prometheus Books

Magick, Mayhem, and Mavericks: The Spirited History of Physical Chemistry. Copyright © 2002 by Cathy Cobb. All rights reserved. No part of this publication may be reproduced, stored in a retrieval system, or transmitted in any form or by any means, digital, electronic, mechanical, photocopying, recording, or otherwise, or conveyed via the Internet or a Web site without prior written permission of the publisher, except in the case of brief quotations embodied in critical articles and reviews.

Inquiries should be addressed to
Prometheus Books
59 John Glenn Drive
Amherst, New York 14228–2197
VOICE: 716–691–0133, ext. 207
FAX: 716–564–2711
WWW.PROMETHEUSBOOKS.COM

06 05 04 03 02   5 4 3 2 1

Library of Congress Cataloging-in-Publication Data

Cobb, Cathy.
    Magick, mayhem, and mavericks : the spirited history of physical chemistry / Cathy Cobb.
        p. cm.
    Includes bibliographical references and index.
    ISBN 1–57392–976–X (alk. paper)
    1. Chemistry, Physical and theoretical—History. I. Title.

QD452 .C63 2002
541.3'09—dc21

2002070511

Printed in the United States of America on acid-free paper

Contents

Acknowledgments   9

To My Physical Chemistry Kindred   11

Preface   13

Introduction   17

PART I. ARISTOTLE AND THE ANCIENTS: MATHEMATICS, MOTION, AND MACHINES

Introduction   25
1. The Ancients: Heavenly Geometry   27
2. Aristotle: Touching Earth   35
3. The Arabs: Mohammedan Amalgamation   45
4. Medieval Magick and Renaissance Revival:
   Finding the Force   53

  5. The Merchants: Mathematics of Money
     and Machines of War   63

PART II. THE EUROPEAN SCIENTIFIC REVOLUTION: THE PASSIONATE PURSUIT OF ORDER

Introduction   73
  6. Mathematics: The First Wave   80
  7. Physics: The First Wave   92
  8. Chemistry: The First Wave   101
  9. Mathematics and Physics: Wave after Wave   111
 10. Lavoisier: The Chemical Tsunami   121

PART III. THE FIRST ATOMIC WARS: DALTON TO PERRIN

Introduction   133
 11. Dalton's Diminutive Friends: An Atomic Theory   138
 12. Thermodynamics: The Warmth and How It Spreads   147
 13. Gnats in Sunbeams: Kinetic Theory of Gases   157
 14. Statistical Mechanics: Maxwell's Demon
     and Boltzmann's Dream   166
 15. Thermochemistry: Willard Gibbs
     and the Quiet Insurrection   177
 16. Atoms or Not? The Test of Jean Perrin   186

PART IV. PHYSICS AND CHEMISTRY COME TO LIGHT: SPECTROSCOPY TO QUANTUM MECHANICS

Introduction   195
 17. Spectroscopy: Rainbows from the Sun   198
 18. Electromagnetism: The Nature of the Sun   207

19. Atomic Structure: Pieces and Puzzles   217
20. The Quantum Revolution: Little Bits of Stuff   228
21. Quantum Riddle: When Is a Particle Not a Particle?
      When It's a Wave.   239

PART V. THE FLOURISH OF THE PHYSICAL CHEMIST: AFFINITY AND FORCE

Introduction   247
22. Rebels and Radicals: Quantum Chemistry in the
      United States   250
23. System and Symmetry: The Further Developments
      of Quantum Chemistry   262
24. The Ionists and Affinity: *Physikalische Chemie*   275
25. Intermolecular Forces: The Tie That Binds   285
26. Heirs to the Ionists: Physical Chemistry
      Crosses the Ocean   295
27. Reaction Rates: Physical Chemistry Keeps Pace   306

PART VI. THE FRUITS OF THEIR LABORS: THE PRODUCT OF PHYSCIAL CHEMISTY

Introduction   319
28. Physical Chemistry and Physiology:
      Not So Strange Bedfellows   325
29. Nonlinear Dynamics: Physical Chemists Make Waves   337
30. Nanotechnology: How Low Can You Go?   348
31. Extreme Quantum: The Questions Keep Coming   357

Epilogue   369

Notes   375

Selected Bibliography   399

Index   403

Acknowledgments

I gratefully acknowledge the excellent work and steady hand of my editor, Linda Greenspan Regan. I also wish to thank my parents, Dr. and Mrs. Clarence Cobb, and my husband, Monty Fetterolf, who read, corrected, and commented on every chapter. I thank my children, Mathew, Benjamin, and Daniel, for their patience and love. I wish to acknowledge the generous assistance of Howard Rasmussen, instructor of history at Aiken Preparatory School; Thuy Nguyen, chemistry laboratory instructor at Augusta State University; Barbara Johnson; Shirley Macintosh; and the Augusta State University library staff. I wish to extend my heartfelt gratitude to Dr. Keith Laidler for reading what I thought was a nearly finalized draft and pointing out many errors. Any remaining errors are, of course, my responsibility alone. I also wish to acknowledge and thank my students, in particular Frank O. Onyemauwa, who helped me, inspired me, and above all, questioned me and held me to my task. Peace.

I have attempted to give you a glimpse . . . of what there may be of soul in chemistry. But it may have been in vain. Perchance the chemist is already damned and the guardian of the pearly gates had decreed that of all the black arts, chemistry is the blackest. But if the chemist has lost his soul, he will not have lost his courage and as he descends into the inferno, sees the rows of glowing furnaces and sniffs the homey fumes of brimstone, he will call out:

Asmodeus, hand me a test tube.

G. N. Lewis, physical chemist, circa 1920

To My Physical Chemistry Kindred

In looking back over this, my *magnum opus*, what strikes me is not how much there is, but how much has been omitted. Writing a history of a science that makes the story accessible to scientist and non-scientist alike is a tricky proposition. For me, it meant leaving out the lives of many fascinating people for the sake of continuity and glossing over many intricate details when explaining concepts with words, rather than our accustomed equations. I do not feel too guilty as there are several wonderful histories of physical chemistry that have been written without these flaws (please see the bibliography and my endnotes). I gratefully used these sources throughout this writing. However, if I have offended anyone by my omissions, please accept my apologies, and if there are inaccuracies, please let me know. On an encouraging note, a student who read part of my manuscript commented, "I would have done better in thermodynamics if I had read this first." On hearing this, I felt hope of achieving my goal.

Preface

When most people think of physics they think of black holes, big bangs, and other astronomical events. While the physics of our celestial universe is decidedly astounding, there is another universe, right here on Earth, that is every bit as amazing. This universe embodies crushing forces and vast vacuums and energies sufficient to perpetuate or destroy life. This universe contains particles that zip around at incredible speeds or swim lazily in elaborate formation: colliding, merging, glowing, and swirling in an intricately choreographed ballet. This universe undergoes ordered pulsations and chaotic explosions. It exhibits astonishing symmetry and breathtaking beauty. It is the universe of atoms, molecules, and the interactions between them, and the physics that explores this diminutive world is as fascinating as the physics that probes the stars.

This is the physics of chemistry—physical chemistry—and its goal is to tease out and tame the forces responsible for the structure and reactions of matter, be it in skyscrapers, enzymes, or stars.

Physical chemistry began with attempts to understand reactions wielded by sorcerers and alchemists and progressed by the use of mathematics, physics, imagination, and perseverance to comprehend and control what was previously attributed to magic. In the process, physical chemists have discerned the composition of planets and stars without leaving Earth. Physical chemists have learned how to predict if a reaction will fizzle or flame, heat or cool, explode or fail—before the beakers are poured together. They have developed three-dimensional models of biological materials and used these models to design drugs. Physical chemists have worked on the billionth of a meter scale in the curious world of nanotechnology and investigated the intricate interplay of interactions that lead to zebra stripes and the beating of the human heart.

In these pages, I hope to paint a picture of the splendor and complexities of physical chemistry as clearly as I can. As an instructor of physical chemistry and an active researcher in the field for many years, I believe the concepts of physical chemistry can be rendered comprehensible by analogy and example. I also hope to convey in the stories of the heroes and heroines of physical chemistry—the mavericks who ran outside the herd, the renegades who took the path less traveled, the visionaries who saw connections that others could not—the singular liberation of their minds. It is a story of those who dared to perform forbidden arts in their quest for understanding and suffered the consequences of their daring. It is a story of madmen and -women who drowned in paralyzing depression one moment and exploded with stellar creativity the next. It is a story of rebels, recluses, and misanthropes . . . but it is the very eccentricity of these pioneering champions that allowed them to see beyond the obvious and defend their conclusions in the face of derision.

In this history you will meet those who knowingly endangered their health for the sake of understanding and those who were attacked by their peers for dissenting ideas. You also will meet those who flirted with fraud to further their fame or based their conclusions on the flimsiest of grounds. You will meet those who merely struggled to understand, but were martyred by wars or extinguished by mobs who could not tolerate a freethinking mind. You will meet rulebreakers and soothsayers, prodigies, swindlers, and nerds—but with one thing in common: the compulsion to go where others will not and the need to know, whatever the cost.

The story of physical chemistry is a celebration of the free spirit, the glory of the unabashedly odd. Physical chemistry is about listening to everyone, questioning everything, learning from everywhere, and synthesizing it all, to create your own—unique—magic.

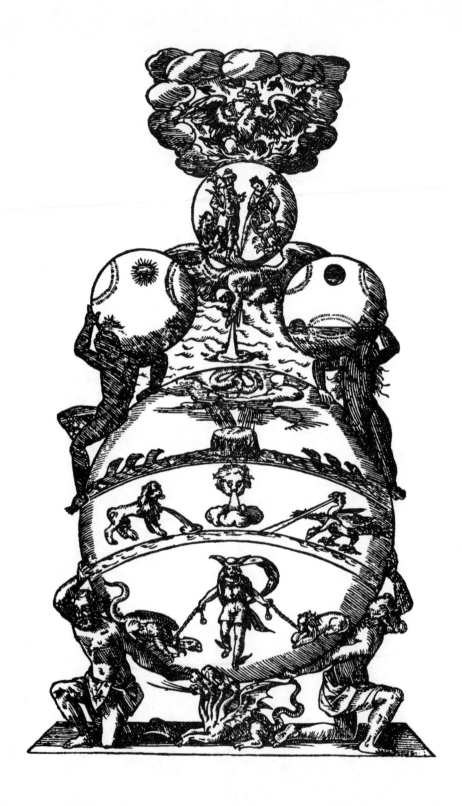

Introduction

. . . physical chemistry is not just a branch but it is the flower of the tree.
Wilhelm Ostwald, circa 1890

Asked about chemistry, most people envision boiling beakers and fuming flasks or recall vague memories of a class in organic, general, or biological chemistry, but few are familiar with physical chemistry. Yet the theory of physical chemistry is what enables us to determine the power of explosions, the composition of clouds, and the structures of particles millions of times smaller than the point of a pin. We use physical chemistry to decide the metabolism of dinosaurs, the makeup of stars, and the rates of reactions within the living cell. The goal of physical chemistry is to develop mathematical models that explain and predict the structures of molecules and the forces that govern the behavior of chemicals in reactions. Through the results of physical chemistry, a synthetic chemist can calculate, before pouring the potions together, what should react—or overreact—and how much heat should be required or produced: knowledge that could spell the difference between rocket liftoff or launch pad disaster. Using the tools of physical chemistry, a biochemist can build a model of a biological molecule, such as the enzyme the HIV virus uses to copy itself, and predict the shape of a molecule that would block its function and prevent HIV replication.

But for all its importance, physical chemistry is far from a household phrase. One reason is that physical chemistry is a mathematical science whose product is equations, and as such rarely finds itself in the popular press. Another reason is that physical chemistry, compared to the venerable fields of organic and inorganic chemistry, is a relatively young science. Classical physics—that is, pre-quantum and pre-relativity physics (cannonball physics)—began with the observations of Aristotle, advanced with the mathematics of Newton, and became a solid systematic science by the beginning of World War I. Chemistry—as a repertoire of materials and reactions—began with the need for medicines, grew with the industries of civilization, and became an impressive body of knowledge by World War I. In contrast, at this time, physical chemistry was a still nascent science. The job of physical chemists is to explain the structure of molecules and the behavior of chemical reactions, but at that time they did not have a complete description of even the simplest atom.

Chemistry's lag behind physics in the development of mathematical models has caused some commentators to regard chemistry as a somewhat humbler science. Anthony Standen, author of *Science Is a Sacred Cow*, captures the sentiment: "Chemists are, on the whole, like physicists, only 'less so.' . . . Much of what they do is . . . related to cooking, instead of a true science."[1] But this assessment is not accurate. The delayed development of mathematical models for chemistry has nothing to do with the relative mental capabilities of the physicist as compared to the chemist nor is it a question of the validity of either science. Rather it has more to do with the subject matter under study. Whereas the classical physicists were observing the motions of planets and apples falling off of trees, the chemists were trying to understand the behavior of matter, which is composed of atoms. And the atom is a very nebulous thing.

An atom cannot be seen with the naked eye, not even with an

ordinary microscope. Nowadays physical chemists can "see" atoms with sophisticated microscopes that use electrons instead of light—but even then it is debated as to what exactly is being "seen." To get an idea of the size of an atom, consider a single grain of sand. It's very small, perhaps only a half of a millimeter wide, but you can still see it. How many silicon atoms would it take, lined up side by side, to span a grain of sand? About 2 million.

In addition to dealing with subjects that are too small to see, physical chemists also have to contend with the fact that these atoms are always in motion. Atoms and molecules fly across the room like gas-phase bullets or squiggle and squirm like a can of worms in even the coldest of condensed materials. In the 1600s, when Johannes Kepler published his brilliant conclusions about the motion of the planets, he did so based on the observations of an unhappy hermit, Tycho Brahe. Brahe was able to provide ample data about the motion of the planets because he could go out and make a sighting; come in, have dinner, then make another sighting; go to sleep, wake up, then make another sighting—all without worrying that the planets might have moved out of view. The chemists' object of study, however, can zoom around at hundreds of meters per second and suffer some billions of collisions in the same amount of time. They have such low mass that any use of light to measure their position knocks them off course. Imagine how far astronomy would have progressed if the moon wavered in its orbit every time someone pointed a telescope at it. Atoms and molecules don't even have the courtesy to have nice rigid boundaries or to bounce off each other in simple elastic rebounds. Atoms and molecules are sticky. They tend to deform or explode on collision in all manner of ways. At least the projectiles of cannonball physics have the good grace not to turn into new materials when they roll into one another.

Given the above considerations, the physical chemists may perhaps be forgiven their slower progress—or even be applauded for making any progress at all. Only after the foundations of mathematics, physics, and chemistry had been laid could the physical chemist begin to build. Before scientists could develop the physics of things you can't see, they had to develop the physics of things you can see, which is why our history of physical chemistry includes much that is the history of physics, too. Clear concepts of motion, force, light, and action at a distance were essential to the development of physical chemistry. To develop a theory of chemistry, the physical chemist needed an infrastructure of mathematics; therefore, the history of mathematics factors into the history for physical chemistry, too. Physical chemists use the power of statistics, algebra, geometry, trigonometry, calculus, group theory, differential equations, vector analysis, and more. But physical chemistry is based on physical phenomena, and these phenomena have an existence independent of the mathematics used to describe them. In short, the physics of chemistry can be described through analogy and example, without the explicit need for equations, an objective that will be pursued throughout these pages.

But perhaps the highest hurdles for physical chemistry, and indeed all science, can be summed up in one unwieldy word—*paradigm*. This unfortunate word has the power to repel by its very form and sound but it is the one word that suits our purpose. A paradigm is an organizing principle or pattern—the basic assumption on which the rest of a theory is built—and as such is very useful. Unfortunately, once a paradigm is embraced, it is exceedingly hard to unseat. The philosophers of Pisa still believed different masses fell at different rates—Aristotle's dictum—even after Galileo offered powerfully logical arguments against it. Mendeleev, the originator of the periodic law on which the periodic table is based, refused to accept the existence of the noble gases—helium, neon, argon, krypton, radon—because they did not fit his original version of the table. Berthelot, a major player in the development of thermodynamics and reaction rates, destroyed others' careers because he could not accept the idea of spontaneous reactions that did not give off heat, effectively denying the existence of modern portable "cold packs." Einstein, in reference to Max Born's interpretation of matter waves as probability waves, said, "God does not play dice with the universe." But, to challenge another paradigm, given the success of Born's theory, perhaps She does.

Because the subject of physical chemistry is the study of objects you can't always see, undergoing changes you can't always measure, and behaving in ways you can't always predict, physical chemistry has a long history of encountering contentious resistance to the paradigm shifts it has proposed. Therefore the study of the history of physical chemistry is instructive because it cautions us to examine our ideas carefully and take care that we do not impose previously held beliefs on newly acquired data. History teaches that we must always question our assumptions and even authority. When it comes to science, skepticism is healthy. We only make progress when we see beyond the present paradigm. To appreciate or participate in the advancement of science, we must be prepared to revise, sometimes drastically, our perception of the universe. In many ways, the history of physical chemistry is an ideal venue for examining paradigm shifts, interwoven, as it is, with the history of chemistry, physics, and mathematics.

Accordingly, in Part I of this book, "Aristotle and the Ancients," we begin by examining how the Greeks, building on Egyptian and Babylonian mathematics, achieved an inaugural paradigm shift: They separated science from religion. One Greek in particular, Aristotle, parented another major paradigm shift: He put forth the idea that the input of the senses is more important than imagination in deducing the nature of the world. Unfortunately, the senses can deceive, as any optical illusion will demonstrate, and many of Aristotle's conclusions were flawed. However, his efforts constituted an important beginning and an essential step forward in systematic thought.

The Greek society was eventually conquered by the Romans, but their philosophy survived in the Byzantine civilization and became assimilated into the collected wisdom of the Arab Empire. Thus, the Greek tradition was transmitted to the budding European cultures via the Arab Empire and Byzantines fleeing the Turks. The Europeans, inspired by the ideas of Aristotle and the knowledge of the Arabs, absorbed both, but did not immediately digest them because they were locked in a Judeo–Christian tradition of not questioning authority. Even in this stifling atmosphere, however, a curious breed of magicians would sow seeds of skepticism that would be the earmark of the physical chemist, and it is through these magicians that the physical chemists trace their intellectual heritage. Unlike the alchemists who laboriously sifted through ancient knowledge in the search for false gold, the magician merrily collected the curious, both old and new—magnets, mirrors, potions, and explosives—and examined them with an attitude of incredulity, an eye to tomfoolery, and a realization that all that was to be known was not necessarily in the past, but could be in the future. During the European Renaissance change also came from another quarter—the artisans—as personified by Leonardo da Vinci, a painter and mechanical engineer by trade. The work of these artisans, driven as it was to build better armaments as well as create beauty, inspired a closer look at the conclusions of Aristotle.

In Part II of this book, "The European Scientific Revolution," which spans the period 1500 to 1800, we find that criticizing Aristotle became fashionable in Europe in the early part of the era, even among the respective popes. However, the Protestant Reformation and the resulting Thirty Years War sobered the Roman Catholic Church into reasserting its authority in scholastic matters through the terrible tool of the Inquisition. As a result, Aristotle was reinstated. But the Church could do little, as could anything else, to truly impede scientific progress at this point. Experimental refutations of Aristotle's conclusions inspired the radical notion that even the input of the senses had to be subjected to empirical test. Hence, the important ideas of motion and force became quantified and more clearly understood. At the culmination of the Scientific Revolution, the vague notion of elements evolved from Aristotle's earth, air, fire, water into Lavoisier's distinct materials with quantifiable properties such as mass and reactivity.

At the beginning of Part III, "The First Atomic Wars," we will see how this clarified notion of elements led Dalton to propose an atomic theory of matter built on a solid foundation of experimental findings, thus setting the stage for a true science of physical chemistry. We will then see how the clarification of the concept of heat, coupled historically with the European Industrial Revolution, was joined to the atomic theory of matter through statistical mechanics. This new theory brought about a momentous change in the way science perceived the world: No longer was nature a predictable, clockwork machine that could be quantified with infinite precision; now science had to admit some knowledge was only accessible through statistical probabilities. By using this approach, in the shadow of an impending world war and surrounded by fervent controversy, the question of the atomic nature of matter was, for the moment, put to rest.

In Part IV, "Physics and Chemistry Come to Light," we will see how new discoveries in the nature of light, electricity, and structure fed into Dalton's atomic theory and culminated in the quantum theory of energy and matter, the repercussions of which are still being wrestled with today. Building on the results of quantum mechanics, in Part V, "The Flourish of the Physical Chemist," we will see how in the space of a hundred years there came a veritable explosion of understanding. From the mid-1800s to the mid-1900s, physical chemistry came into its own. Scientists around the world tackled the problems of quantum chemistry, heat, chemical equilibrium, solution properties, electrochemistry, and the rates of chemical reactions. The importance of intermolecular forces and the role of relativity were tackled. The puzzle of chemical affinity, that elusive, capricious attraction between chemicals, was subjected to mathematical modeling with unpredicted, yet pleasing, results.

Then, in Part VI, "The Fruits of Their Labors," we will examine some of the marvelous applications of physical chemistry, many of which, such as physiology, may seem surprisingly far flung from either physics or chemistry. But modern physical chemists take pride in the fact that they are able to contribute to all branches of science and that they do not fall into any particular fold. They are an independent lot, as their calling demands. They have to be able to bridge the disciplines and converse in many camps. A good physical chemist is more than a student of mathematics, chemistry, and physics: Modern physical chemists need to be students of biology, geology, and astronomy because this is where their skills apply.

Hopefully this survey of the history of physical chemistry can help demystify the discipline and make the precepts more understandable, but it must be admitted that nothing can completely simplify the science. However, it is my hope, that in light of this history, the reader may come to see the beauty amid the complexity. To produce *The Last Supper*, da Vinci used the mathematics of perspective, the chemistry of pigment and paint, and the physics of the scaffold. To produce his transition state theory, Henry Eyring used the physics of quantum mechanics, the chemistry of molecular structure, and the mathematics of statistical mechanics—but when he was finished, for the world and all time, he, like Leonardo, had created a work of art.

Part I

Aristotle and the Ancients

MATHEMATICS, MOTION, AND MACHINES

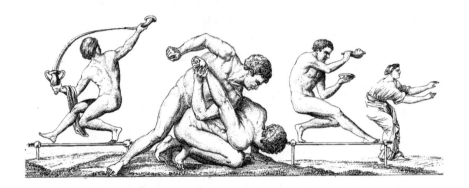

INTRODUCTION

Plato is dear to me, but dearer still is truth.

Aristotle, circa 350 B.C.E.

George Santayana, American philosopher and poet, said, "Those who cannot remember the past are condemned to repeat it."[1] While this statement may ring true for political history, how can it apply to the history of science? Surely we do not have to reinvent the telescope or rediscover the laws of motion just because we forget the circumstances of their introduction. Yet there are lessons we risk losing if we neglect our scientific past. The first is that science is evolving. It is not static. It is not complete. We must remember that perfectly intelligent people, in culturally rich societies, have constructed reasonable and intuitive models that later proved to be incorrect. Therefore, if we are to progress, we must not become complacent. We must continually question our assumptions, no matter how authoritatively stated. The second is that scientific insight knows no boundaries. It may come from the rich or the poor, the free or the slave, the East or the West. Scientific inspiration may come to anyone, anywhere, who is willing to learn.

And so we begin with the ancient Greeks. The Greeks assembled some of the earliest mathematical models for the physical world and introduced the notion that mathematics is an end unto itself; that is, mathematics for the sake of mathematics. Aristotle took Greek philosophy a step further: He showed that the ethereal philosophies of the ancients could also be applied to the natural world. We will proceed to examine how the Arabs helped introduce Aristotle's ideas, as well as new ideas in medicine and mathematics, to budding European societies.

The early Europeans often venerated the knowledge of the ancients to the denigration of their own efforts because they had seen the wonders of the ancient world. For example, the earliest Europeans built from bricks discarded by the Romans because they did not know how to make their own.[2] For the medieval Europeans, algebra from the Arabs was like a bolt from the blue. We will see how this spell began to be broken by the skepticism of the European medieval magicians. The mechanical engineers of the European Renaissance will then provide two essential ingredients for the European Scientific Revolution: the ability to measure and the belief that things should be measured. Early European science was not just a matter of learning about natural phenomena, it was a matter of learning that they could learn.

Chapter 1

The

Ancients

HEAVENLY GEOMETRY

Whatever we Greeks receive we improve and perfect.

Plato, circa 400 B.C.E.

The ancient Greeks were excellent architects, perceptive politicians, and brilliant philosophers—but a bit muddle-minded when it came to the physical sciences. So it may seem rather incongruous, in this history of physical chemistry, to focus so much on the Greeks. But we will, and for several good reasons. While they added little in the way of actual scientific fact, they established important models and patterns of thinking in mathematics, logic, and astronomy. Because of certain political and social events, the precedents established by the Greeks would strongly influence, for better or for worse, the development of physical chemistry for 2,000 years. The names of the Greek philosophers, in particular Aristotle, would be invoked by academics, magicians, scholars, and sorcerers. Finally, it may be that the ancient Greeks were among the first in the Western world to divorce science from religion and attempt to explain observed facts without immediate reliance on divine intervention. Thus cast out of the garden of religion and superstition, they originated new modes of thinking about mechanics, mathematics, logic, and astronomy.

Though the importance of logic and mathematics to chemistry may be self-evident today, one may still ask how astronomy is related to chemistry. We belabored the point that atoms are so tiny they cannot be perceived by ordinary means, so how could the study of objects hundreds of quadrillions of times larger than atoms be important to physical chemistry? One reason is that atoms, like poles of a magnet, sometimes attract, sometimes repel. Understanding such forces that act at a distance is fundamental to the progression of physical chemistry. One of the first such forces to be described was that of gravity, and the understanding of gravity came from astronomy.

But the understanding of gravity did not come from Greek astronomy because the ancient Greeks had it wrong, which points out another connection between ancient Greek astronomy and physical chemistry—models. The Greek astronomical models, which had the heavenly bodies traveling in perfect circles and mostly around Earth, were quite ingenious and very reasonable, given the instrumentation of the day, but they were wrong. Therefore consideration of their work helps us acquire the healthy skepticism necessary for rigorous science. Models are only as good as the methods used to test them. Nice, neat models always need to be regarded with a critical eye. In physical science, there is always the danger of falling in love with the system rather than the nature that inspired it. We shall see instances of this fallacy throughout our story of physical chemistry. Perhaps the most important gift to be gleaned from the work of the ancient Greeks is that it is always possible to devise a perfectly plausible model that is wrong—a memory that will haunt the efforts of physical chemists through their history and one that cautions them now.

Nonetheless, the ancient Greeks also provided a way over the hurtle they created: mathematics, in particular geometry and trigonometry. Geometry and trigonometry are essential tools for studying the motion of particles acted on by forces, which includes atoms as well as planets. It was not just the machinery of mathematics the Greeks provided, they initiated a new way of thinking about mathematics, too. To the Greeks mathematics was an art, an enterprise for its own sake, justified by its beauty alone.

The people we call the ancient Greeks are actually the Dorian Greeks who came to Greece from places unknown around 1000 B.C.E.[1] By 500 B.C.E. they had also founded cities in Asia Minor, on the Mediterranean coast, on islands such as Crete, and in southern Italy and Sicily. The theory has been put forth that in some of the more far-flung outposts, where they experienced contact with other civilizations and were out of contact with their own, the Greeks learned the habit of thinking for themselves and relying less on tradition.[2] Etymological evidence indicates that the word "philosophy" was an African word with roots found in southern as well as northern African languages[3] and there is good evidence that Thales (640–546 B.C.E.), a representative of the earliest Greek philosophers, was educated in Africa.[4] Though later Greek philosophers would eschew work in any form and deem commerce demeaning, Thales was an entrepreneur who made his fortune by cornering the market in olive oil presses. Beginning the tradition that would become the classic custom of the Greek philosopher, he took up the study of mathematics, particularly the geometry he had learned from the Egyptians.

Geometry, the mathematics of points, lines, angles, shapes, and the relationships between them, is an elegant discipline. Although the Egyptians and Babylonians had knowledge of some rudimentary relationships, such as multiplying the length of two sides of a rectangular space to find the number of tiles needed to cover its area, they used this knowledge to help them construct buildings and mark off fields, but they did not attempt to extend their methods in ways that were not immediately useful. The ancient Greeks thought differently. The Greeks became enthralled with the game of mathematics. Although this approach may at first appear pointless to the engineers among us, this new way of thinking was to have a tremendous impact on the sciences. Time and again, pure mathematics—the pursuit of mathematics for its own sake—would result in mathematical tools that would eventually find important, pragmatic applications. When the later European mathematician, Reimann, had initially presented his geometry based on curves instead of straight lines, it had no practical application whatsoever.[5] But when Einstein needed a mathematical system with which to frame his theory of general relativity, he found and used Reimann's geometry. Accordingly, as we will see throughout this history, the ancient Greeks will have geometry in place when Galileo needs it; Noether, Galois, and their ilk will have the mathematics of symmetry and group theory in place when the quantum chemists need it; the mathematics of statistics will be in place when Boltzmann needs it; and Newton will have calculus in place when everyone needs it.

For the early Greeks, discoveries in mathematics were everywhere for the finding. Because of its beauty and symmetry, a ninety-degree angle—the "L" angle made by the sides of a perfect square—became a focus of interest. Thales found that extending two lines from the arc of a semicircle to the ends of its base will always form a right angle. He is said to have sacrificed a bull in thankfulness to the gods when he discovered this one.

A right triangle is a triangle that contains one right angle. The side opposite the right angle is called the hypotenuse. The famous theorem that states that the sum of the squares of the sides of a right triangle will equal the square of the hypotenuse is called the Pythagorean theorem after Pythagoras, who was probably Thales' student. The Babylonians already knew of this relationship, but the followers of Pythagoras, the Pythagoreans, were the first to offer a proof. Proof by logic was another way in which Greek mathematics differed from Babylonian and Egyptian mathematics. The beautiful, logical proofs had such elegance that the Greeks became enamored of proofs and decided that every proposition in every field could only be accepted if it had a logical proof.

The Greeks employed three main systems of proof: induction, deduction, and *reducio ad absurdum*. The first, induction, is the development of a working premise based on a series of observations or an obvious truth and is the fundamental logic that we use every day to navigate our way through life. For instance, using induction, we might decide that "all dogs bark" based on the fact that all the dogs we have observed to date were able to bark. Thales, based on his observations on triangles, might have used induction to decide "all triangles with two equal angles have two equal sides."

The next type of proof, deduction, is based on the premises found by induction. In deduction, the conclusion must follow from the premises. For instance,

> Premise: All dogs bark.
> Premise: That animal is a dog.
> Conclusion: That animal barks.

Or for Thales

> Premise: All triangles with two equal angles have two equal sides.
> Premise: That triangle has two equal angles.
> Conclusion: That triangle has two equal sides.

From these examples, perhaps, it can be seen why this elegant logic works well for mathematics, but is limited in it usefulness in the physical sciences. A biologist or a veterinarian may have seen the flaw in the first argument: While it is true that the vast majority of dogs bark, there is one breed of dog that does not: The basenji. In an inductive proof, all it takes is one contradictory example and the premise is false. So the first premise is false, and for the ancient Greek logicians, this would have been a showstopper. They would have thrown out the whole argument. But notice how the conclusion might still be useful to a biologist or veterinarian for purposes of species identification: Barking is one of the usual properties of a dog. So while rigorous deductive proof is useful in physical science, it cannot be regarded as the only goal as it was in ancient Greece. The premises must be weighed for their usefulness and conclusions tested for their validity.

Reducio ad absurdum is the process of rejecting a premise by following its consequences through to an absurd conclusion. Take again the premise "all dogs bark." From that premise we would have to conclude that the neighbor's dog, dead some several years and buried in the back yard, is capable of sitting up and barking—an obvious absurdity. We will see Galileo employ *reducio ad absurdum* in his analysis of motion.

The Greeks delighted in the bounty and beauty of their mathematics. After Thales and Pythagoras followed many others. The *Elements* of Euclid (circa 300 B.C.E.) is such a refined compendium of geometry that it was used until the mid-1800s as the standard text in geometry.[6]

It is easy to imagine how the Greeks, empowered as they were by their elegant system of geometry, must have believed that they could solve any mystery, including the mystery of the stars. And, in truth, they nearly did. Trigonometry, as a set of relationships between sides and angles of a triangle, was also known to the Babylonians and Egyptians and had been used by them to calculate distances and areas for the laying out of fields and the construction of buildings. The Greeks pointed their triangles to the heavens.

Of the Greeks who attempted astronomical calculations, some of the most notable are Hipparchus (190–120 B.C.E.) and Ptolemy (100–170 C.E.)—not to be confused with the Egyptian dynasty of the same name. Although Hipparchus and Ptolemy flourished a few hundred years after Plato (circa 428–348 B.C.E.), they were still strongly influenced by his thinking.[7]

Plato, whose name is usually regarded as synonymous with Greek philosophy, has not been mentioned so far in this history, but his influence cannot be understated. Plato is usually remembered for metaphysics rather than mathematics and as being diametrically opposed to the empiricist. He denigrated the input of the senses and argued that tangible objects were but imitations of the ideal object. Therefore, to Plato, the planets and stars were perfect orbs in the sky and their movements were in perfect circles.[8]

Again, we dwell on this ancient astronomy because of the importance of the lesson it teaches: The ancient Greeks were so skilled in geometry and trigonometry that they were able to piece together a very believable astronomical theory that accounted for their observations, though based on completely erroneous assumptions. Later, when other theories such as the atomic model for matter were proposed, its defenders would point to how well the theory agreed with observation and its detractors would point to the errors of the ancient Greeks.

The great moral of ancient Greek astronomy is that a perfectly plausible theory can be developed that explains and predicts—a sun in orbit around Earth explains sunrise and predicts sunset—but is still incorrect. Therefore a careful scientist learns to use phrases such as "there is good evidence for" or "the data point to" but to eschew any statement that rings of "this proves." Science must be viewed as a work in progress: A new premise must generate predictions that can be tested, then the results of the test must be used for refinement of the premise, which leads to refinement of prediction, which leads to refinement of test—which leads to refinement, refinement, refinement—ever closer but never at the end. It is easy to imagine why this approach did not sit well with the Greeks, who had cut their teeth on the beautifully exact systems of geometry and trigonometry.

In mathematics, the definition of line and angle can be agreed upon and a proof based on these definitions is an end, a conclusion. In the sciences, the natural world cannot be "defined," but has to be accepted as is. In physical chemistry, as in all science, our models are just that—models—not the real thing. A floor plan of a house is a useful model and can be used quite conveniently to predict how many people can live comfortably in the house. But people live in the actual house, not the floor plan, and there is a world of difference. In the sciences the word "prove" is shunned in deference to the realization that a final "proof" is only as good as the premises on which it is based.

For example, an observer on Earth, such as an ancient Greek astronomer, would see Mars trace an apparent zigzag pattern across sky. Our modern explanation is that Earth and Mars are on two different orbits around the sun, so we are like a child on a carousel watching a waiting parent: sometimes the parent appears to approach and sometimes the parent appears to recede. However, the Greeks were also able to account for the zigzag pattern with another geometrical construct: *epicycles*, or circular orbits within circular orbits. To understand why epicycles would produce a zigzag pattern, envision a ship far out at sea but still visible on the horizon. If this ship were steaming in small circles, to a watcher on shore it would appear to be going back and forth on the horizon. Two different models can thus explain the same result and given the measurements the Greeks were able to make at the time, there was no way to distinguish between the two. Eventually, with advances in instrumentation such as the telescope, enough evidence would be found for rejecting the epicycles, but it would still take a considerable effort to overcome intellectual inertia and unseat the notion for good.

In summary, the ancient Greeks did science a great favor by adopting and cultivating mathematics as a pursuit in its own right. They also advanced the cause of systematic thought by insisting that their truths derive from logic. Some of their conclusions were flawed, but taught a valuable lesson in the use, and misuse, of models. And as we mentioned above, in addition to mathematics, logic, and astronomy, the ancient Greeks also considered mechanics: the study of forces and their effect on motion. Because of the importance of force and motion to physical chemistry, mechanics, along with its progenitor, Aristotle, will be allotted a chapter unto itself.

Chapter 2

Aristotle

. . . that which is set . . . in motion, in turn . . . excites motion in another portion . . .

Aristotle, circa 350 B.C.E.

A ristotle is a major player in the history of physical chemistry, and it is unfortunate that much of the time we are forced to regard his influence in a negative sense. While it is true that he established erroneous conceptions of motion and matter that had to be uprooted before the current notions could thrive, Western science is forever indebted to Aristotle for a critical change in the manner in which explanations were sought: Aristotle, unlike his mentor Plato, placed a high regard on the input of the senses—sight, smell, taste, sound, and texture—and established a new legitimacy for observation. Misguidedly, he, as many before and after, found his system so satisfying that he attempted to extend it to cover the whole of experience, which could not be done. And through a series of political, social, and natural circumstances, the input of Aristotle's senses would grow to have

Fig. 1. An energetic teacher and writer, Aristotle taught while restlessly roaming around the grounds of his Lyceum, so much so that his teachings came to be known as the peripatetic (walking or strolling) school of philosophy. *Heck's Pictorial Archive of Military Science, Geography, and History,* © 1994 by Dover Publications, Inc.

disproportionate importance in Western science. There would be, however, those who understood what Aristotle had actually intended. Galileo, while refuting Aristotle's concept of the heavens, is said to have remarked that he wished that Aristotle could have seen his experiments. He believed Aristotle would have approved.[1]

Busts of Aristotle (384–322 B.C.E.) reveal that he had a pleasant demeanor, and though he is said to have talked with a lisp, one can see how his countenance commanded respect. Aristotle was raised in the relative outback of the Macedonian court while his father served as physician to Philip II, the father of the future Alexander the Great.[2] Aristotle no doubt learned something of medicine from his father because he retained a fascination with biology and physiology throughout his life. At the age of seventeen he was sent to Plato's Academy in Athens, where he remained for the next twenty years. On the death of Plato, he wandered through the various Greek colonies, traveling eventually to Asia Minor and marrying. After five years abroad, Aristotle returned to Macedonia and became tutor to the thirteen-year-old child who was to become Alexander the Great. This relationship would serve as one of the most important factors in securing Aristotle's place in history.[3]

After his years as a tutor, Aristotle returned to Athens, where he founded a school of his own, the Lyceum. Unlike Plato's Academy, which was mainly devoted to social and political philosophy, Aristotle emphasized the study of biology and nature. Aristotle may have written some dialogues for popular dissemination while he was at Plato's Academy, but these have been lost, and the body of Aristotle's work that has come down to us is, in fact, his lecture notes from the Lyceum.[4]

In contrast to the meandering style of the dialogues of those days, the lecture notes are organized, categorized, and illustrated. In the first half of his notes he examines topics such as logic, metaphysics, memory, sleep, dreams, longevity, youth, and old age. The entire second half, however, concerns biology: *History of Animals*. Aristotle's extant lecture notes contain some 700,000 words (as translated into English from the original Greek) and when one considers that 100,000 words make a good-sized novel, this amounts to seven good-sized novels—handwritten in columns, on parchment, in scrolls, complete with anatomical diagrams. As energetic a teacher as a writer, Aristotle taught while restlessly roaming around the grounds of his Lyceum, and his teachings came to be known as the peripatetic (walking or strolling) school of philosophy.

Aristotle was fascinated by the quest for system. Plato, Aristotle's mentor, had expended a great effort to systematize politics, metaphysics, and even the art of living. Euclid would organize geometry by introducing an axiomatic system in which theorems are derived from basic principles. This type of systematization was what Aristotle envisioned for physical science: the synthesis of a set of axioms, based on common sense observations, from which the whole of physical experience could be explained. To this end he employed classification schemes and the system of deductive logic, taking as his premises the input of his senses.

However, conclusions based solely on human sensation can be incorrect. A classic example is the moon. When we see the full moon low on the horizon, we see an apparently gigantic orb. But later in the night, when we look up at the moon in the sky, we see what appears to be a much smaller object. Has the moon shrunk? Not according to modern science. Nowadays we explain this perception as an optical illusion: The mind compares the size of the moon to the objects around it, and on the horizon the moon appears huge with respect to the trees and buildings by it. High in the sky there are no similar reference points, so the moon appears smaller. But in Aristotle's world there were no optical illusions. What was perceived was real.

Aristotle should not be blamed for his reliance on the input of the senses. The human mind has adapted for survival, not science, and the mind interprets its perceptions without conscious control, often for the better. Most babies, placed on the edge of a glass-covered precipice, will not venture out onto the glass, even when tempting objects are placed on the glass or a parent stands on the glass and coaxes.[5] The baby perceives a danger and will not be convinced it is not there.

Plato suggested a more flexible system for arriving at premises: the dialectic method—a method of inquiry that utilizes a constant questioning of assumptions. The dialectic method is a complex and difficult approach, but it has the advantage of being open ended: A final conclusion may never be reached, but the theory can continuously be refined. But Aristotle rejected the dialectic method just as Plato rejected sensory intelligence.

Our contemporary approach to science is a refined and extended combination of the Aristotelian and Platonic systems, which has come to be called the scientific method. In the scientific method one first observes then forms a hypothesis, or guess, to explain the observations. The hypothesis is tested by systematic experiment and, based on the results of the tests, the hypothesis is refined and the cycle repeated. This method incorporates the Aristotelian approach of proceeding from observation or experiment, but also allows for continual refining of the interpretation.

The open-endness of the modern scientific method would not have sat well with Aristotle. Aristotle's passion for orderliness caused him to reject what he could not immediately see. Through his observations he arrived at his infamously erroneous theories, notably the four-element theory, the belief in the continuous nature of matter, his erroneous notions in mechanics, and his support of the geocentric theory of the solar system.

Aristotle's four-element theory is found in his compendium *Meteorology*, described by his biographer L. Minio-Paluello as being the corpus of Aristotle's "physical chemistry."[6] The contents of *Meteorology* indicate how closely he would have identified with the goal of physical chemistry as practiced today. For example, modern chemistry begins with the classification of materials by different physical or chemical properties. A physical property is a characteristic that a material will have while standing alone on the shelf, such as its color or the temperature at which it melts. A chemical property requires the admission of another material to be measured, for instance, flammability is a chemical property and the tendency for flammability can only be measured by providing oxygen and heat. Aristotle spent a good deal of time delineating the physical and chemical characteristics of materials. He also described how materials could form from what he took to be the four basic elements: fire, air, water, and earth.

Although the four-element theory did not originate with Aristotle, he embraced it on the grounds of sensory perception. The ancients could see that many substances had a watery component, such as blood, so they reasoned that water had to be part of their makeup. If they could see bubbles escaping when a liquid was boiled, they assumed air was part of the composition; a reasonable assumption. When wood is burning, the products—fire, smoke, sap, and ash—clearly resemble the four elements. Aristotle also used observation to support his contention that materials are continuous, not particulate (that is, atomic), in nature. He could look very closely at materials such as water and not see any separate particles; therefore, matter to Aristotle was plainly continuous—so said his senses. However, it is not in this arena that Aristotle had the most influence. It was Aristotle's explanation of motion that was the most difficult to derail.

Motion and its causes and effects were a focus of curiosity for the ancients and remain items of interest today. The reasons are practical as well as esthetic: The design of machines requires an understanding of motion as does the explanation of the flight of cannonballs and the energy of atoms. Aristotle again based his notions of motion on observation. He observed that an object moved when he pushed it and eventually stopped moving when he didn't. To explain continued movement, he postulated that when he initiated the motion in the object he also initiated a movement in the air: The air rushed into the spot where the object had been to prevent a vacuum and this movement of air perpetuated the motion when he removed his hand. Just as rationally, Aristotle concluded that heavier objects fall faster than light ones because he observed a feather fall more slowly than a rock. Nowadays we attribute this difference in the rate of falling to air resistance, and we know that a feather and a rock will fall at the same rate in a vacuum, but this notion still *feels* counterintuitive because it runs against our common experience. To Aristotle's credit it must be noted that it took until about 1600 for Galileo to prove him wrong, until about 1700 for Newton to make a general statement of the law of gravity, until about 1900 for Einstein to provide an improved theoretical explanation for gravity, and we still don't know what exactly gravity is.[7] But, ironically, part of the cause for the delay was the success of Aristotle and his emphatic, authoritative statement of what his senses told him had to be true. It was also Aristotle's insistence on sticking with the sensory that led to his most celebrated error: the geocentric model of the solar system.

Actually, much of what the ancient Greeks knew about astronomy was correct and cleverly deduced. The spherical shape of the Earth was immediately obvious to the ancient navigators who noted that the positions of the stars relative to the horizon changed as they sailed north or south. And anyone who has witnessed a lunar eclipse has seen that the shadow of the Earth on the moon is curved, as had been reported by Aristotle. Eclipses also provided evidence that the moon is smaller than the sun: When the moon passes between the Earth and the sun the moon appears to cover the sun. If the moon were larger or the same size as the sun, it would appear to swallow the sun. (This principle can be demonstrated by moving one's thumb between one's eyes and a distant object larger, then smaller, than the thumb.)

The model of the solar system that Aristotle adopted was based on the notions of Eudoxus of Cnidus (408–355 B.C.E.).[8] In this model Earth is in the center of the universe. To Aristotle, the observation that all things fall toward Earth made sense if one assumed Earth was the center of everything. The motion of the sun was explained by locating it on a great sphere that rotates around Earth. The stars and planets, likewise, occupied rotating spheres in this model. Aristotle elaborated on the theory by adding more spheres until the total was fifty-five. Aristotle had heard of heliocentric (sun-centered) models for our solar system, but he rejected these for a geocentric (Earth-centered) model again on the grounds of what he could observe with his own eyes. He could see the sun rise and set. If the sun were at the center of the solar system and the Earth were rotating around it, then the only explanation for day and night would be that the Earth were spinning around. But if the Earth were spinning around, Aristotle reasoned, then an object tossed into the air would not return to the same spot. Ergo, the Earth does not spin and the sun is not the center of the solar system.

Again, in all fairness to Aristotle, it must be recognized that although the notion was commonly accepted much sooner, it took until the 1700s for direct evidence to be found for the movement of the Earth around the sun: In 1729 James Bradley, after years of exacting observation and recording keeping, found a slight movement in the perceived position of very distant stars, in a direction opposite of what would have been predicted by rotation of the Earth.[9] He was able to show through careful calculations that the effect—the aberration of starlight—was due to the movement of the Earth around the sun.

Shortly after the time of Aristotle, Aristarchus (310–230 B.C.E.) proposed a sun-centered solar system. He was a brilliant astronomer and in his treatise *On the Size and Distances of the Sun and the Moon* he showed how measuring the angle between the quarter moon and the sun could be used to get at the ratio of the distances between the moon and the sun.[10] Using this ratio and other clever triangulations involving the shadows of the Earth and the moon during solar and lunar eclipses, he was able to measure the relative sizes of the Earth, the moon, and the sun. Because another philosopher, Eratosthenes, had measured the circumference of the Earth by the shadow cast by the sun at two different locations and triangulation, Aristarchus was then able to estimate the absolute size of the sun and the moon. But Aristarchus had no real proof for his proposition that the sun was in the center of the solar system, and because his model, based on circular orbits, did not explain phenomena such as the unequal length of the seasons, he was not very confident of his theory. His speculations were side-railed in favor of Aristotle's geocentric theory for the next 1,500 years.

The geocentric theory, as with many of Aristotle's ideas, persisted because it *seemed* correct. Aristotle's theories required only the senses for verification. Aristotle's lecture notes read as an emphatic statement of fact, with none of the questioning tone of the dialogues. In addition to the intuitive appeal of his arguments, another factor that conspired for ascendancy of Aristotle and his ultimate acceptance as a preeminent authority in medieval Europe was the legitimate success of some of his compatriots such as Euclid and a physician called Galen.

We have already mentioned the work of Euclid and his clear and concise compendium of geometry. Given the power of geometry to calculate lengths, areas, and volumes of complicated figures, the methods of Euclid's *Elements* must have seemed like manna to the ancient farmers and builders. Similarly, in the ancient world where people died routinely and in agony from infection, dental decay, and dysentery, Galen's medicines and methods, though flawed by today's standards, must have seemed like miracles.

Galen learned medicine while treating gladiators, and in his trials with gladiators, he determined that more gladiators recovered if wine was poured on their wounds. Wine contains alcohol, which is a disinfectant, so no doubt he had a higher healing rate than physicians who did not disinfect. Galen also learned that diet, exercise, bathing, and massage were good prescriptions for a number of maladies, but because he knew he could sell medicines many times but advice only once, he usually recommended these regiments along with an herb or two.[11] Many of the herbs he used had potent physiological effects such as being laxatives and stimulants. Galen also learned that the pulse was a good indication of the emotional state, so in a manner reminiscent of a fortune teller, he is said to have held the patients' hands while questioning them about the source of their discomfort. The reputation of Galen and the success of some of his cures made his writings much valued, as were those of Euclid. As Aristotle was rediscovered by the Europeans, his writings gained some regard by simple association with the works of Euclid and Galen.

There were political reasons for the persistence of Aristotle, too, namely the success of Alexander the Great. Since his tutelage, Alexander had admired the Athenian Greeks, but he was a cosmopolitan conqueror and admired things Persian and African as well. He attempted to develop a homogenous civilization under his rule, but after he killed Aristotle's nephew, Callisthenes, for lack of obedience (Callisthenes had been dispatched by Aristotle to travel with Alexander and gather specimens), Alexander was remorseful for the offense to his mentor's family and tried to make up for it by building Greek cultural centers and Greek libraries. He promoted Aristotle's ideas in his Hellenized world, which extended like a scorpion with its pincers around the eastern Mediterranean and its bloated torso at the Hindu Kush.

The survival of Aristotle's ideas may have had to do with climate as well as politics and academics: The papyrus plant did not grow in Persia, so the peoples of the East had to use parchment for a writing material, which is more costly and less convenient, but lasts longer. Papyrus scrolls such as the ones Aristotle (or his students) used to record his lectures would have decayed in less than a hundred years, but the parchment of Persia on which they were copied was more durable. The writings of Euclid and Galen as well as Aristotle came to find themselves on Persian parchment because the Greek and Macedonian civilizations were already aging at Aristotle's time and the conquerors soon became the conquered. Rome subdued Macedonia about 200 B.C.E. and in 150 B.C.E. the Greeks were defeated, about the same time as Egypt succumbed. The Romans first suppressed the new Christian religion that was born in their eastern territories, but then in 313 C.E. legalized it. Christians, so long the persecuted, became the persecutors. They attacked and destroyed the ancient institutions in sporadic riots. In one such outbreak they killed, literally tearing limb from limb, one of the last Greek philosophers, Hypatia, renowned for her mathematical skills. Other scholars took heed, fled to Syria and the East, and began to copy Aristotle on parchment. So it was that the teachings of Aristotle survived to eventually be transported to Europe.

Among the important contributions of the ancient Greeks was certainly a "heads up" warning to future thinkers: Science must be skeptical. Data must be subject to test and verification. Acceptable theories must explain *and* predict. But another was a "leg up" from Aristotle. After Aristotle, Western philosophers were liberated to dissect and study the natural world as well as the ideal and to use their eyes and ears to do so. While it may be true that his legacy was, for physical science, imperfect—input from the senses needs to be verified—the importance of his contribution is incalculable.

Chapter 3

The Arabs

I . . . think that never blows so red, the Rose as where some . . . Caesar bled.
Omar Khayyám, circa 1000,
The Rubáiyát, as rendered by Edward Fitzgerald

When two metals mix, the result is called an alloy—except when one of the metals is mercury—in which case the mixture is called an amalgam. Why the special name for alloys of mercury? Because mercury is an unusual metal with unusual properties. To begin with, mercury is a liquid at room temperature, which is not true for any other metal. Mercury can also combine with an amazing range of other metals and does so at room temperature. Gold, for instance, is a mercury sponge, as anyone whose golden ring has accidentally touched mercury can tell you. The mercury will quickly seep into the gold and turn the surface from golden to silvery in appearance.

It is the combining power of mercury that we allude to when we speak of Mohammedan amalgamation: The Islamic Civilization, after its inception in 622 C.E. (year 1 on the Islamic calendar) spread with

remarkable speed, assimilating along the way the accumulated knowledge of the Greeks, Hindus, Romans, Chinese, Egyptians, and East
and West Africans. The causes of this phenomenon were manifold,
including the Muslims' ability in trade, their practice of tolerance,
and their love of the good life.

The Muslims were geographically well situated for trade through
overland routes with the east, land and ocean routes with Africa, and,
once established in Spain, land routes to the north. Travel was the
Arabic way. The Arabs were originally nomadic and the acceptance of
Islam brought with it five pillars of the faith, one of which is a pilgrimage to Mecca at least once during a lifetime. Muslim trade routes
linked Asia, Africa, and Europe.

In their travels, whether for conquest, pilgrimage, or trade, the
Arabs retained a societal trait that subtly influenced the sciences: tolerance. Of course this tolerance may have been born of practicality
rather than idealism—only non-Muslims were taxed so it was good to
have a lot of them around. It also must be noted that the tolerance
was not universal: The Islamic empire was able to tolerate within its
borders other monotheists such as Zoroastrians, Jews, or Christians,
but they put to the sword the polytheists.[1] However even if they were
able to tolerate only some, they were certainly able to trade with, and
learn from, everyone. To understand the importance of this ecu-
menical attitude, recall those portions of the now defunct Roman
Empire that insisted on retaining their Roman identity: the Byzantine
Empire and the Holy Roman Empire. They were also saddled with
Roman numerals—I, II, III, IV, V, and so on—for use in daily calculations. For anyone who has ever attempted simple arithmetic calculations with Roman numerals, let alone something like multiplication or
division, the disadvantage of this awkward notation is apparent. However, the Arabs could shift and bend in the direction that best suited
their needs; therefore, the adaptation of the numeral system of the
Hindus—1, 2, 3, 4, 5, etc.—including the all-important concept of
zero, gave them considerable advantage in commerce and navigation.

As a result of the affluence brought by trade, coupled with the
tolerant acceptance of new ideas from different cultures, the Islamic
peoples developed an appreciation for the good life, which included

silks from China, gold from Africa, jewels from India, and books from Alexandria. The Muslims introduced paper to the Western world, which they had learned to manufacture from Chinese captive laborers. Paper helped them copy and keep books and produce their own. The caliphs were wealthy and enthusiastic patrons of knowledge and, under their encouragement or command, works from other cultures were translated and scholarship flourished. But the Muslims were more than a conduit. They augmented and enhanced what they absorbed. They chose their own pragmatic approach over Hindu aestheticism or Greek elitism and generated their own advances. Three clearly identifiable areas of Arab contribution important to the development of physical chemistry are mathematics, instrumentation, and that storehouse of chemical information, alchemy.

Let us look first at mathematics and start at zero. Positional notation means simply that the value of the digit depends on its position in the number. For instance, in the number 250 the 2 is in the hundreds position and we interpret the number 250 as having two hundreds, five tens, and zero ones. Positional notion was used by the Babylonians in their base-sixty system and by the Romans with their Roman numerals, but the problem was that they had no zero to mark the empty position. So the number 250 would be written as 25_, a two and a five followed by a blank space, but this was long before the day of uniform type and typesetting, so if the blank was a little bit short or a little bit long, 250 could easily be interpreted as 25 or 2,500. The advantage brought about by the zero, and popularized by the Muslims, is apparent.

Perhaps it was the acceptance of this idea of a symbol for nothing that allowed the acceptance of the idea of a symbol for anything, but for whatever reason, another important branch of mathematics that flourished under the Muslims was algebra, a system which derives its name from the Arabic word "al-jabr," or "restoration."[2]

There is some evidence that the Greeks knew the basic idea behind algebra—that a number could be represented by a set of facts about that number. For instance, a person could say, "the number I'm thinking of is four plus two." Then the number is obvious, six. The "set of facts" about the number is the equation; therefore, algebra consists of representing numbers by equations.

Algebra, of course, is capable of doing much more than solving riddles. And it is also able, with the right number of equations, to solve for more than one unknown, which makes algebra a very powerful system. For instance, an ancient Arabic olive merchant may want to determine what to charge for olives and for oil if he has a hundred barrels of olives and two hundred barrels of oil and he wants to make a profit of ten dinars over his original investment in olive trees. A physical chemist may want to know how many liters of ethanol and how many liters of water to mix to have six liters of a 30 percent ethanol mixture, taking into account the fact that the solution shrinks as you mix it. In the hands of the Muslims, with help from the Hindus, algebra became capable of solving such problems and more. They used their intricate system of algebra in commerce, law, metallurgy, and agriculture.

Of all the famous Muslim mathematicians, perhaps the most famous is Omar Khayyám—but he is known more for his poetry than his algebra. Khayyám, who flourished about 1000 C.E.,[3] was an accomplished mathematician and wrote several treatises on the subject. Another famous Arabic mathematician whose name is used every day by mathematicians, physicists, and chemists, though they probably don't know it, is al-Khwarizmi.[4] Al-Khwarizmi's name in its Latinized form has come to mean a procedure for solving a mathematical problem: algorithm.

Al-Khwarizmi, who flourished about 820 C.E., was mathematician for the court for the caliph of Baghdad. Though he did not claim originality for the system of algebra, and made frequent reference to his Hindu sources, it was his books that survived to become important in Europe, and it was the name of one of his books—*Al-jabr wa'l muqábalah*—from which the word algebra (al-jabr) is derived. Al-jabr is generally taken to mean *restore*, or *fix* (in *Don Quixote* the word *algebrista* means bone-setter), and refers to the process wherein the equation is fixed, or simplified, by adding or subtracting terms from both sides. It is interesting to note that al-Khwarizmi, like others of his time, wrote his equations in prose rather than the symbols that pervade modern algebra books.

In addition to algebra, the Arabs made great advances in spherical trigonometry, a three-dimensional trigonometry that could be used for astronomical calculations and navigational instruments such as the astrolabe, an ingenious device that resembled an elaborate handheld sundial. With the astrolabe, navigators could tell the time of day *and* estimate longitude and latitude. Astronomers could determine positions and movements of stars. Another instrument for which the Arabs found extensive use was the balance, though it was used for a more earthly pursuit.

The state of the Arabic art of mass measurement by balance is evident in the writings of al-Khazini, who flourished in the early 1100s.[5] He was a Byzantine eunuch slave who had been well educated by his owner. In a treatise on instruments al-Khazini describes a balance of his own design, which, after employing an algorithm that he also devised, could determine the composition of alloys of two different metals. This ability to determine the purity of metal may have been influential in the development in another pursuit that would eventually be of major importance to the development of physical chemistry: alchemy.

Alchemy, like other words such as alcohol and algebra, is distinctly of Arabic origin because of the use of the article "al." The Arab alchemists came up with some interesting new materials such as the "alcohol that burns," meaning alcohol so pure it would catch fire. They also made sal ammoniac, or ammonium chloride, a material unknown to the Greeks, and better and more concentrated caustic solutions. But as far as their influence on later European alchemy, their myths may have meant more than their knowledge: The European alchemists were so convinced that the Arabs knew the secret of transmutation—the conversion of base metal into gold—that some tried uttering Arabic-like words as part of their experiments. Some of these have come down to us today as the magical words of "alacazam" and "abracadabra." But while alchemy has become most strongly associated with the later European efforts to make gold from base metal, the term originally referred to any chemical pursuit, that is, any effort to find recipes to manufacture materials. Alchemy encompassed all efforts to make stronger acids, purer metals, and better medicines, as well as attempts to make gold. For the Arabs, the efforts to make gold did not dominate their culture. The reasons may include the fact that they were able to detect fake gold on their balances because gold has a different density (a different ratio of mass to volume) than other metals. They also had plenty of gold through trade (in 1324 Mansa Musa, a West African king, distributed so much gold on his pilgrimage to Mecca that the gold market in Cairo fell).[6] Muslims believed efforts to make gold to be avaricious, and, perhaps through the influence of the Chinese, they found the manufacture of medicines to be as important a goal as the making of gold.

The Arabic interest and knowledge of medicine was impressive and this may again have been a function of their amalgamating power. The pre-Islamic Arabs had their own medical tradition kept by the old women of the tribes, but in the Arab Empire, all the traditions of the conquered territories were added. The Arabs conquered for knowledge as well as taxable territories. When one considers some of the known folk remedies that have a verifiable activity—salt, beer, and urine for antiseptics; kaolin clay for an antidiarrheal agent; rhubarb for a purge; willow bark for an analgesic; quinine for fever; and egg yoke for skin lesions—one has to wonder who had the audacity to try the material in the first place! In some cases the information may have come from observing animals and what they ate, but in other cases the discovery may have been sheer serendipity or sheer desperation. So the accumulation of pharmacopoeias had to have been a long, painstaking task, with only a few real treatments coming from any one group. Add to this the fact that not all plants grow in all areas, and it becomes apparent what an integrating force the Arabs were in compiling this knowledge.

Once accumulated, this knowledge became valued by the Europeans. However it may not have been so much the special effectiveness of Arabic medicine as the magnitude of the hole it had to fill. A description of European treatment during the period of the Crusades can be found in the writings of Usama ibn-Munqidn.[7] According to the writer, a Frankish noble requested medical assistance from a Muslim prince and the prince obliged. When the doctor arrived he was shown two patients, a knight with a leg abscess and a woman suffering from a condition described only as "dryness." The Arab doctor used a dressing to bring the abscess to a head and let it drain after it burst, a procedure that would have allowed the sore to heal if it had been allowed to proceed. But a Frankish doctor, who had also been called but arrived later, advised amputation saying that the patient would die without it. The patient did die—during the amputation.

The Arabic doctor had ordered a diet that included fresh vegetables for the woman, but this was discontinued on the advice of the Frankish doctor and her hair was cut instead. The "dryness" increased so the Frankish doctor declared that a demon had entered her body through her skull, and he initiated a horrific surgical procedure to extract it. The woman also died. The Arab doctor then returned home, his services no longer needed. However, it must be noted that the Arab physicians learned of several mineral and vegetable substances with medicinal activity from the Franks and entered them into their pharmacopoeia with due acknowledgment to the Europeans.

Islamic influence on Europe began with the occupation of Spain in 700 and reached the height of its vigor in the 900s. Contact increased through the eight major crusades held between 1095 and 1291. But the eventual devastation of the Islamic Empire came from the east, not the west: the invading Mongol hordes of Genghis Khan. Along the trade routes established by the Muslims, the Europeans came in contact with the Mongols.

Although there were Arab geographies and a book by a Spanish Jew, Benjamin of Tudela, explaining how to reach China by trade route, there was little interest in China by Christian Europe of the 1100s. What did interest the Europeans were the letters from the apocryphal Prester John who told of a rich Christian king in the Orient. Popes dispatched envoys in hope of making contact with this king and converting the Mongols they encountered along the way. Though these efforts met with little success, the new interest in the Orient gave a boost to traders, such as Marco Polo, who set down his account in a book written from prison. Although Marco Polo wrote his book some years after his adventures, it is not the fanciful account sometimes portrayed. He did not describe any monstrous races—dog-headed men, pygmies fighting with cranes, giants, and so forth. These are products of Greek and medieval European legend. At the end of the 1400s copyists added them as illustrations even though they were not found in the text. Because of this confusion, the bulk of Polo's account was discounted. For the development of physical chemistry, this may have had a retarding effect. As we will see in our discussion of thermodynamics, the Industrial Revolution in Europe, fueled by coal, had a major impact on physical chemical theory. This interest might have come about sooner if Marco Polo's account of "a kind of black stone existing in beds in the mountains which they dig out and burn like firewood" had been believed.[8]

However, despite some unfortunate disbelief and some misplaced credulity, Arab and Eastern knowledge began to infiltrate into European culture. Stimulated by the Muslims and the Mongols, Europe underwent a commercial revival beginning about 1300. The Europeans eventually embraced the new mathematics and Aristotle's teachings, but did so only selectively. For instance, Aristotle rejected the notion of a disembodied soul because it could not be perceived by the senses, but the concept of soul was basic to medieval Christianity. They readily accepted Aristotle's Earth-centered universe, but in this Aristotle may have been the messenger, not the message. It is very probable that Christian Europe would have rejected a sun-centered universe, even if Aristotle had proffered one, because the Earth-centered universe fit their concept of Earth being God's first creation. Through the addition of commentaries, Christian authorities adjusted Aristotle's notions to fit their own. This adjustment led to an ironic situation: Aristotle, who held so strongly that the senses should be believed, and the Arabs, who imported, experimented with, and added to information from other cultures, were now made the authorities in a rigid, closed system of *Scholasticism,* which did not allow for any deviation from the established norm. There were, however, those who found a way around the system. They claimed the authority of Aristotle for everything they wished to try, and then while the scholars were searching their volumes for confirmation or contradiction, the natural magicians had their fun.

Chapter 4

Medieval Magick and Renaissance Revival

. . . here are two sorts of Magick; the one is infamous, and unhappy, . . . and consists of incantations and wicked curiosity; and this is called Sorcery; . . . The other Magick is natural; which all excellent wise men do admit and embrace, and worship with great applause . . .
Giambattista della Porta, circa 1600

The span of time that includes the European Middle Ages (from around the fifth century to the early 1300s) and the Renaissance (from the early 1300s to the late 1500s), is a complex period of cultural history in which tradition did much to shape science. In the next two chapters we will examine primarily the lives of two Renaissance representatives, a magician and a mechanical engineer. The magician will serve as our backward-looking mirror of the Middle Ages and the engineer will be our telescope to view the future. It is perhaps a commentary on the importance of the individual, as much as tradition, to note that our forward-looking engineer actually predates our magician by about a hundred years. Our magician is Giambattista della Porta, and Leonardo da Vinci is our engineer.

To appreciate the work of della Porta and da Vinci it is necessary to understand the context of their times. It was an age of dichotomies: Practical technology coexisted with sorcery and courtroom logic was used to justify the madness of witchhunts. The forces that drove the developments of physics and chemistry were manifold—Arabs, plagues, Inquisitions, Scholastics, and homegrown technology—and these forces converged over lands where pagan traditions, including Celtic tales of elves, fairies, and Merlins, were fused with the Christian religion. The introduction of the compass catapulted Europe into an age of exploration, which meant more merriment for the mix, and it also introduced the fascinating phenomenon of magnetism that would eventually have extensive repercussions in physical chemistry.

J. D. Bernal, in *A History of Classical Physics*,[1] put forth the theory that magnetism was discovered by the ancient Chinese. In a ritual for determining the location for their graves, they spun a precious object and dug where it pointed when it came to rest. Bernal points out that the future residents knew that they would spend a great deal of time at these locations, so the object spun was made with the most precious material available: gold, jade, or lodestone, a naturally magnetic variety of iron ore. Someone must have noticed that while gold and jade reported random locations, the lodestone would have a disproportionate number of people buried in the north (or south, depending on how the spinner was read). It wasn't long before this curious locator was put to use in navigation. Better navigational tools increased European contact with non-Europeans, which triggered an economic revival, which brought with it a revival in learning and innovation. But increased contact had a downside: the spread of the bubonic plague or Black Death, characterized by hemorrhages under the skin that turned black and enlarged lymph nodes, or buboes, hence bubonic. The great plague and its aftershocks instilled a type of paranoia in the people that was intensified by the uneasiness of the popes in the face of the Protestant Reformation of the early 1500s. In an attempt to regain control of religion, the Catholic Church insti-

Fig. 2. Demon of the Plague. From H. von Gersdorf's *Feldtbuch der Wundarzney*, printed by Johann Schott, Strassburg, 1540. *Devils, Demons, and Witchcraft*, © 1971 by Dover Publications, Inc.

tuted a systematized effort to seek out and discourage freethinkers. This effort, known as the Inquisition, turned against all perceived enemies of the faith, including Jews, Muslims, worldly clerics or laity, and even midwives. Among midwives there were no doubt some who did dispense poisons or "inheritance powders" for a fee or indulge in self-administered hallucinogenics.[2] Yet, the majority of them practiced a benevolent art in which they prescribed herbal extracts of legitimate use such as ergot and belladonna to ease pain and aid in the recovery from childbirth or inhibit premature contractions.[3] But at a time when the sufferings of labor were considered just punishment for Original Sin, to ease such pain was considered heresy.[4] So interestingly, while the European Middle Ages is sometimes perceived of as a steady progression from darkness to enlightenment, in this one regard, at least, the progression was backwards: Persecution of midwife witches accelerated after the 1300s and peaked about the time of the Scientific Revolution of the 1600s.[5]

However, the establishment of universities was also very much a function of the Catholic Church and it was Church scholars who eventually integrated Aristotelian logic and Arabic learning into European thinking, though in their own uniquely medieval way. One of the basic tenets of the monotheistic Judeo-Christian-Islamic tradition is revelation, the acceptance of truth on faith, and revelation was the perspective from which the medieval mind viewed the world. But the powerful pull of Greek inductive and deductive logic could not be ignored. Initially medieval theologians wrestled with the reconciliation of reason with revelation. Eventually they were in the main convinced, through the efforts of Scholastic philosophers such as Thomas Aquinas,[6] that the two were not necessarily at odds. Yet when they assimilated both the logic and the science of the Greeks and Arabs, they used Greek logic to derive proofs of the existence of God while accepting the science verbatim, as a sort of scientific revelation, without any effort to explore or modify it.

If the Middle Ages can be said to be a mix of forward steps and backward steps, then Scholasticism can be viewed as a side step. Scholasticism sought to reintegrate religion into science whereas the Greeks had endeavored to separate the two, yet Scholasticism allowed for examination of Aristotelian and Arabian ideas, which inched things forward. The Scholastics wrote extensive commentaries to expound Aristotle in a manner that was in accord with Christian precepts and these commentaries came to carry the same, or more, weight as the works from which they derived. In the process of adaptation of Aristotelian thought to a system based on faith, Aristotle himself became an icon of sorts; it became near heresy to deny the contents of any of the approved commentaries written about him.

But one innovation was sufficient to break the cycle of commentary on commentary, and that was the European reinvention of printing with movable type in 1450 (the Chinese already had a similar technique).[7] The effect was that the translated ancient Greek and Roman classics were made more available and this availability of the translations made the commentaries less useful. People started reading the translations for themselves. It also opened the door for a revival of an old trade: writing scientific treatises for profit. Previously writers needed a patron such as a caliph or other ruler to subsidize their writing, but now there opened the possibility of writing independently for profit. A new type of personality took on the challenge of science: not philosophers of means or employees of a monarch, but artisans and people in the trades. Our magician and our engineer were both tradespeople: Leonardo da Vinci, born 1452, was primarily a painter by trade and Giambattista della Porta, born 1535, was a writer.

Della Porta was born in a villa outside Naples sometime in the late 1500s.[8] He had some tutors, but was mostly self-taught, a natural consequence of being a second son in an age of primogeniture. He probably received some support from his family because he was wealthy enough to join learned societies and to marry. However the additional income brought in by the sale of his books was apparently welcome. Although he wrote books on optics, mechanics of water and steam, mathematics, military fortifications, and distillation, his best known work by far is his compilation *Natural Magick: Wherein Are Set Forth All the Riches and Delights of the Natural Sciences.*[9]

In this age of dichotomies, magic was one more. Della Porta was one of several of his age to practice what was termed "natural magic," as opposed to sorcery.[10] Sorcery, in the vein of Faustus, was based on

Fig. 3. Title page from della Porta's *Natural Magick.*

incantations and ritual. Those who indulged in natural magic, on the other hand, were interested in collecting tricks and curious marvels.[11] To these magicians, magic meant wondrous effects and illusions, but illusions cleverly and knowingly wrought. Della Porta took care to divorce himself from sorcery by pointing out that his magic didn't involve incantations. It was not that he disbelieved in witchcraft—in his book he offers guards against enchantments—he just rejected it as unproductive for him. He did realize, however, that much that was called sorcery was just trickery and in *Natural Magick* he counsels how to recognize deceit. To della Porta a magician was a "skillful workman" who knew how to extract tinctures and perform distillations and was versed in the "mathematical arts" such as astronomy.

Della Porta published two versions of *Natural Magick:* a four-book version in 1558 and an expanded twenty-book edition in 1589, where a "book" is more the size of what we would call a chapter today. The books were compilations of what he termed "secrets" that he had been collecting since he was fifteen and covered an enormous range of practical, everyday problems with a smattering of the extraordinary. In some ways it reads like a medieval compendium of household hints, and in others, it reads like an experimental treatise on physical phenomena.

In many ways, della Porta is a perfect symbol for the Middle Ages. His writings alternate innocently and unabashedly from incredibly gullible to nearly scientific. He quotes remedies, methods, and other particulars with reference only to Aristotle and other ancients for proof of their veracity, while at other times he is careful to add that he has tried the particular method for himself. For instance, he asserts that shellfish are generated out of mud, the hardness of their shells preventing any kind of commingling, and quotes Aristotle as his source. But he also reports a method for catching fish by their attraction to light and explains how he personally tested this idea by constructing a metal tube with glass windows sealed with pitch into which he lowered a candle. He swings from a humane concern for human comfort in his prescription of medicines to a horrifying cruelty in his attitude toward beasts, stating nonchalantly that the most tender meat is procured from an animal that has been tortured before death. He provides detailed diagrams and a table of measurements to show how to construct a parabolic lens for optimum focusing power, but gives recipes for chemical explosives as a bit of this and a bit of that, then says what more may be added if it doesn't work.

He covers practical problems such as animal husbandry, fruit production and grafting, food preservation, egg incubation, cookery, hunting, fishing, tempering steel, and medicines and remedies for teeth, eyes, conception, the pox, and the plague—but also the stuff of amusements such as beauty secrets for women, the crafting of musical instruments, and the extraction of perfumes. He exhibits his own sense of wonder when covering topics such as gunpowder, burning waters, invisible writing, eyeglasses, water clocks, alchemy, counterfeiting precious stones, and magnetism in lodestone. He indulges his own sense of fun when describing a trick cup that delivers water instead of wine (and functions on the difference in density between water and wine), providing methods for turning a woman's face green, or whimsically naming his last catch-all book of miscellaneous magic "The Chaos."

Notable is his omission of human birthing. Although he did not hesitate to recommend medicines for sleep, poisons and antidotes, and antiseptics including wine and salt—and he was familiar with the female anatomy (he recommends methods for tightening the vaginal canal after childbirth and blistering the cervix so that a woman will bleed on intercourse, indicating virginity, and whores "can scarce be known from maids")[12]—he mentions no medicines for childbirth. Undoubtedly this reticence was due to his desire to be dissociated from sorcery and thus midwives. But he did not manage to escape completely. The Inquisition examined him, but did not torture or try him. They satisfied themselves by only banning publication of his works from 1592 until 1598.

Also notable is the cursory coverage of alchemy. Although medieval chemistry is commonly associated with alchemy, della Porta devotes only ten chapters to alchemy as opposed to the fifty-six chapters devoted to magnetism and lodestone and the thirty chapters devoted to cosmetics. However, in this way della Porta is more representative of developments in the Middle Ages that unknowingly affected physical chemistry than were the European alchemists. Although European alchemy does play a part in a comprehensive history of chemistry, in a history of physical chemistry it constitutes a diversion of the stream. Concisely viewed, alchemy had as its goal the transformation of materials, that is, transmutation, primarily with the objective of making gold from base metal and medicines from almost anything. The problem was that gold can't be made from base metal (with perhaps the exception of transformations in a nuclear reactor) but most certainly not by ordinary chemical means. Moreover, the systematic development of medicines requires an understanding of disease, which did not yet exist. So although the Arabic alchemists made much practical progress, the European alchemists were on a fool's errand, and with the exception of a couple of happy accidents, did not advance science very much.

Della Porta was fascinated with fire and self-ignition and describes trials in which he ignites a paper by holding a candle close, but not touching, and an oil fire that will extinguish through smothering but will ignite again when exposed to air. He repeatedly writes that these observations are "very strange" and "wondrous," but he does not seem capable of formatting his questions in a precise manner that could be addressed by experimentation. He plays with vacuums and compressed air by making trick drinking cups and pop guns. When he describes his experiences with these phenomena he does so with a sense of fun and wonder, but not with an eye to improving them or deriving any general principles from them. Della Porta was trapped in an age when technology was presenting questions, but the prevailing philosophy left him without the means to answer them.

Della Porta was aware of the magnifying properties of lenticular crystals (double convex lenses shaped like a pair of parentheses) and he did some experiments with optics, but fell short of inventing the telescope that would later be critical in the Scientific Revolution. He was interested in mirrors, but he played more than experimented with the distorted images caused by warped, funhouse-type mirrors, and in true della Porta style, he became fascinated with multiple mirror reflections and tried to arrange them so he could see to infinity. Della Porta saw all the wondrous potential of his magic but was unaware of the careful, repetitious experimental work that would be necessary to understand it. In da Vinci, our engineer, we will find more confident steps toward experimentation, but they will still fall short of the revolution yet to come.

Chapter 5

The Merchants

Speculators about perpetual motion . . . take your place with the seekers after gold!

Leonardo da Vinci, circa 1500

The European Middle Ages are often envisioned as a technologically impoverished period, with sewage systems consisting of chamber pots dumped into streets and irrigation systems powered by human feet. But by the time of the Renaissance, the Europeans had a respectable amount of technology available to them. They understood simple machines—levers, pulleys, screws, and wheels—and medieval Europeans integrated these simple machines into devices such as water-powered sawmills, pumps, and mechanisms for drawing wire and performing other routine manufacturing tasks. Although they did not understand the cause-and-effect principles of mechanics, the thousand years or so of the Middle Ages were long enough to arrive at these technologies through trial and error. Toward the end of the Middle Ages, the loss of human power to the plague and wars made machinery—and the principles of mechanics—increasingly important. So by the time of the Renaissance it could be said that the forerunners of physical chemistry, experimental chemistry and physics, were conceived, but yet unborn. However, the midwife, mathematics, was by then an old crone.

Europeans were aware of the mathematical achievements of the

Arabs and the Greeks from about the year 1000, but much of the learning remained an exotic novelty. The educated Europeans—the priests, monks, nuns, and rabbis—could calculate the area of a field with Roman numerals and the date of Passover or Easter, which seemed sufficient. The economic revival, however, brought on by the new navigational tools, generated the need for rapid, complex computations such as currency exchange, pricing within an acceptable profit margin, interest calculations, and the division of proceeds in a partnership. No longer could Europe wait for the knowledge of mathematics to filter in. Knowledge of mathematics was sought out and adapted to fit the uniquely European need.

One such person specifically sent to learn new mathematics was Fibonacci (otherwise known as Leonardo of Pisa), born in 1170 to the Pisan merchant family of Bonacci.[1] Around 1192 his father took him to Algeria, where his father directed a Pisan trading colony. Fibonacci used the nickname Bigollo, which meant bum or loafer, which may have suited this son of a wealthy merchant at one time. Once in Africa, however, Leonardo bloomed. He sought out and received instruction in the new Indian numerals and algebra. He developed his skills to the extent that his father began sending him on trade missions to Egypt, Syria, and Byzantium. When he turned thirty, Fibonacci returned to Pisa and spent the rest of his life writing books on mathematics for merchants.

Fibonacci's books are very practical—he gives many examples and offers methods for checking the answers—as well as algorithms for computing interest, profits, measurements, and currency conversion. But he did more than apply the mathematics to commerce. In a manner reminiscent of the ancient Greeks, he was a creative mathematician. He is credited with a series of numbers—called the Fibonacci series—that are generated starting with 1 and then adding the last two numbers to get the next number in the series: 1, 1, 2, 3, 5, 8, 13, 21, and so forth. Ratios of adjacent Fibonacci numbers are found in nature in the 21:34 and 5:8 opposing-spiral pattern in the center of a daisy flower and the pinecone, respectively.[2] Less whimsically, Fibonacci also found that he was able to solve certain bookkeeping problems only if he "conceded that the first man had a debt."[3] In conceding this debt, Fibonacci introduced European merchants and mathematicians to the concept of negative numbers. Later, during the Renaissance, another mathematician, Luca Pacioli, would also write and sell books on mathematics for merchants. Pacioli's *Summa de arithmetica*, written in 1494, included many of the topics of Fibonacci's book as well as a treatise on the new method of double-entry bookkeeping referred to as the Venetian method. This method revolutionized bookkeeping and the concept of debit and credit would take root in other disciplines as well. In the mid-1700s we will see a revolution in chemistry based on a balance sheet—a mass balance sheet—filled out by a businessman by the name of Lavoisier.

Pacioli, like Fibonacci, could not ignore the pull of recreational mathematics. His particular interest was in the geometry of solid bodies. This interest happened to overlap with the interests of another tradesman with whom he became a good friend: a painter, Leonardo da Vinci.

The friendship between Pacioli and da Vinci was an older man's friendship. Da Vinci was forty-seven years old and Pacioli was fifty-two when da Vinci illustrated one of Pacioli's books on solid geometry. Beyond their common interest in geometry, they also seemed to share a similar approach to Renaissance life. For instance, both men were apparently celibate, though some speculate that da Vinci was homosexual. But given da Vinci's own assessment of human reproduction— "The act of procreation . . . [is] so repulsive that if it were not for the beauty of the faces and the adornments of the actors and the pent-up impulse, nature would lose the human species"[4]—he probably was asexual rather than homosexual or heterosexual. Given the risks of an active social life of the times—which included exposure to syphilis as well as the plague—he may have chosen wisely. Pacioli, a friar, apparently made the same choice. But if both men had found peace in some areas of their lives, they were still intellectually restless. Their friendship was founded on da Vinci's desire to learn more about mathematics. As it was, da Vinci, being the illegitimate son of an Italian noble and a peasant, had been taught basic reading and writing, but then was apprenticed to learn a trade instead of being sent to university. The trade that he learned was painting, which was a fortunate choice for humanity, but it may not have been the choice he would have made for himself. From his own words and from the positions for which he applied it appears that he may have had an inclination toward another vocation—mechanical engineering—but his aspirations in this area would be clouded by his deficits in mathematics.

During the Renaissance, however, painting was a lucrative trade. The new funds brought in by commerce had created a wealthy class who enjoyed spending their money on luxuries. For example, the *Mona Lisa* is now believed to be the portrait of the wife of a Florentine merchant. It was generally thought pointless to spend money on investigations of nature because it was believed that all that could be known was already known. There were a few who were swindled into investments in alchemy, but not many, and these ventures were short-lived: Either the swindlers absconded with the investment or were tortured by the impatient patron. Painting gave da Vinci the opportunity to indulge his interests in engineering because painting was considered a trade at that time and painters, considered tradespeople, could be called upon to perform other design tasks such as contriving canals, bridges, and catapults. Aware that there was money enough to invest in art and war—the two major preoccupations of the Renaissance ruling class—da Vinci soon established himself as a valuable commodity. He told potential patrons that he could design machines of war—and paint a bit as well.[5]

The above should then explain why we have placed a painter in this history of physical chemistry. The principles of mechanics—the study of energy, forces, and their effect on bodies—will be central to the development of the theories of physical chemistry. While da Vinci is obviously best known as an artist, he also designed, built, and put into operation many mechanical devices and spent time investigating the principles of mechanics. His patron for sixteen years, Duke Ludovico Sforza—nicknamed the Moor—had him design a heating system for the duchess's bath as well as stage shows employing mechanical devices. Another patron, Cesare Borgia, used da Vinci as a military engineer. Although da Vinci himself did not establish any new mechanical principles, he collected his information on mechanical systems with the intention of compiling one or more books. Unlike della Porta and Fibonacci, however, da Vinci's books were never written—or if they were written they were lost—but we still know much about what he did, and much about the Renaissance in general, because of his wonderful notebooks.

Da Vinci began writing his famous notebooks when he was about thirty. While it is obvious he had every intention of organizing his ideas at some future time, in the notebooks he allowed himself to write or sketch whatever was on his mind at the moment, in a more or less random fashion. He doodled and wrote reminders to himself, drafts of letters, designs for weapons, anatomical drawings, building plans, and grocery lists—the outpourings of a tirelessly energetic mind. In fact it may be because of this intellectual energy that no compilation of his notes came about in his lifetime—he was continuously moving on to new projects, many times at the expense of the completion of the last. Evidence of da Vinci's rapid thought processes can even be found in his handwriting: He habitually wrote backwards, from right to left, in a mirror-image script that can be read by holding the text up to a mirror. It is now assumed that he did not do this for security purposes—with a little practice it is easy to read mirror-image print and when he wanted to keep something secret he wrote in code[6]—but he was left handed and, in a day of quill-and-ink writing, writing from right to left kept him from having to drag his hand through wet ink as he wrote. His rapid thoughts allowed no time for the ink to dry.

Yet, for all the hodge-podge, we will forever be indebted to da Vinci for this undisciplined hopping about from idea to idea because of the wonderfully broad window it gives us into the Renaissance world and what was then known or believed about mechanics. For instance, in da Vinci's time and even much later, engineers were convinced that machines could increase the work done by a person, even infinitely, if the right machine were found. This belief that such perpetual motion machines could be found led to much speculation and many fantastic, fanciful designs. Da Vinci tried his hand at a few, but then realized the pointlessness of the quest. He came to the important conclusion that machines do not generate work but only modify the manner of its application. In other words, the best you can ever hope for is to get out what you put in—and you usually get out less.

For example, you can move a pile of dirt in one load with one huge effort or in many loads with many smaller efforts, but the net amount of work is the same. If you need to move a heavy rock, you can use a lever to do so, and it may seem easier, but you have to apply the force on the lever arm over a much greater distance than the distance that the rock actually moves. Perhaps Archimedes (liberally paraphrased) said, "Give me a lever and a place to stand and I can move the Earth," but he should have also explained that he would have had to transverse the diameter of several Earths to do so. But extending this analysis to all machines was not something easily accomplished with the medieval mentality, so speculations on perpetual motion persisted.

Da Vinci was also unique in that he recognized and took account of friction in his analyses and realized there was no way to defeat this force either. Although a viable theory would not be put forth until a full three centuries later, da Vinci knew, through experience, that friction could be reduced, but not eliminated. Friction has two components: drag due to plowing by surface irregularities as one piece tries to slide over another and the attractions of the surface atoms of one piece for another. This attraction is the "stickiness" of atoms that would have to await a full theory of atomic structure to be even partially understood. But because of this stickiness, any effort to smooth the surface also puts more surface atoms in contact with one another. So even an infinitely smooth surface will experience friction, maybe even more so—there can be a "welding" of one material to another through surface contact.

To improve his machines, da Vinci made attempts to measure the relationship between velocity, time, and distance for an object moving under the influence of a gravitational field, but he lacked the measuring tools and the mathematical skills necessary to accomplish it. This conundrum caught da Vinci and the rest of the Renaissance world: If he had been given a university education he might have had the mathematical skills he needed. Had he been given a university education, however, he might not have tried to make the measurement because the universities were teaching that Aristotle had already stated all that needed to be known about motion. Da Vinci, free from the scholarly community, was able to reject Aristotle at times. One is struck by the crudity of his measurements of both space and time: He used a braccio, which is approximately twenty-four inches, for his smallest unit of length, and he used hourglasses and water clocks to measure time. But though he did not have the means to do the measurement, his methods were on the right track. He understood the value of repeating the experiment with slight, controlled variations of parameters and he understood that errors of sensory observation—optical illusions—could occur. For instance, as an artist, he understood how a perfectly flat canvas could be made to have the illusion of three-dimensional depth: perspective.

Da Vinci was certainly not the only Renaissance artist to employ the techniques of perspective. Da Vinci's own specialty was aerial or atmospheric perspective, that is, the notion that light and dark are in sharper contrast and contours are more defined in objects that are closer than those that are far away. This technique is similar to a perspective technique in use in eastern Asia at approximately the same time. But other Renaissance artists used fully mathematical analyses to achieve their realistic proportions. One of the leading contemporary proponents of perspective was Albrecht Dürer (1471–1528), a German painter who wandered as far as Italy during his journeyman's years of travel.[7] (In those days a journeyman was really expected to journey!) There he became convinced that the new art should be based on mathematics. He methodically constructed the figures for his masterpiece *Adam and Eve* using the instruments of geometry: a compass and a ruler.

This ability to project a three-dimensional object onto a two-dimensional surface would prove essential to the theoretical understanding of chemistry. Molecules and atoms are three-dimensional entities that move in a three-dimensional world. While this may seem self-evident to someone reared on modern concepts, it is important to note that it would not be until the middle of the 1800s that the three-dimensional structure of molecules would be firmly established as a tenet of molecular theory. And, as mentioned above, painters such as da Vinci and Dürer were also concerned with other projec-

Fig. 4. The Four Witches. Engraving by Albrecht Durer, 1497. *Devils, Demons, and Witchcraft,* © 1971 by Dover Publications, Inc.

tions whose analysis would also figure prominently in the theory of chemistry—the projections of cannonballs. The Renaissance interest in cannonballs had recently arisen from a discovery that was old to Asia but new to Europe: gunpowder.

Gunpowder was probably known in China as early as the 800s, but used only for firecrackers. This explosive mixture is a combination of saltpeter, sulfur, and charcoal that on ignition forms gas-phase oxides of the sulfur and carbon from the oxygen of the nitrates of saltpeter. The reaction is so rapid that a container holding the ingredients does not have time to expand gradually and instead explodes. By the time of the Renaissance, Europeans had used gunpowder to power projectiles for several hundred years. Da Vinci himself recorded various recipes for gunpowder for different applications.[8] With the various conflicts of the Renaissance the stakes became higher for the design of new machines of war and more creative applications of gunpowder. The mechanics of objects in motion became critically important when they were needed not just to save the labor of peasants but to preserve the power of kings. And with the advent of mounted cannon on ships, no longer did an enemy ship have to close in for capture; the toy of della Porta—the magnifying lenticular lens—now had the power to alter sea battles and change the course of nations.

So the stage was set for the next paradigm shift: the replacement of Scholasticism with pragmatism. Tradition mattered very little on the battlefield—rulers were more interested in winning wars than paying homage to Aristotle—and soon defiance of Aristotle in the name of scientific progress would become a parlor game for the young bloods of the universities. When these young bloods predicted the trajectories of cannonballs and pointed their telescopes at the stars, the parlor game became the Scientific Revolution.

Fig. 5.
Academie
Laboratory.
The Acad-
emy of
Sciences in
Paris,
Charles
Nicolas
Cochin
(1688–1754)
engraver.
Edgar Fahs
Smith Col-
lection, Uni-
versity of
Pennsyl-
vania
Library.

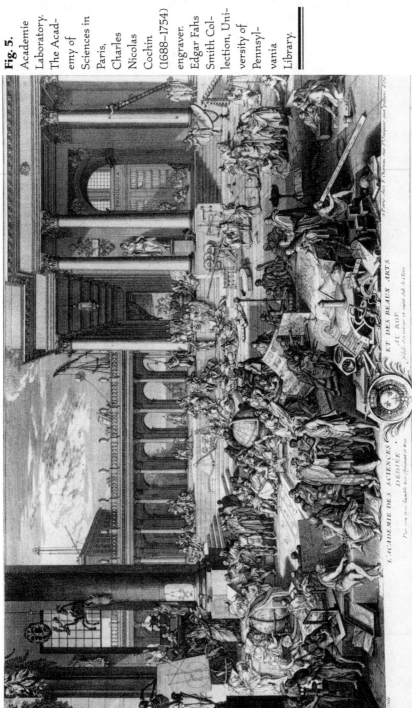

L'ACADEMIE DES SCIENCES ET DES BEAUX ARTS
DEDIEE AU ROY

Part II

The European Scientific Revolution

INTRODUCTION

The miracle of the appropriateness of the language of mathematics for the formulation of the laws of physics is a wonderful gift, which we neither understand nor deserve.

Eugene P. Wigner, circa 1960

Physical chemistry is a mathematical science, but history is not. The Scientific Revolution is generally placed between 1550 to 1700, but in a history of physical chemistry we must extend the time to 1800 before completely consigning Aristotle to the history books. Physical chemistry is a composite science and revolutionary milestones are found from 1550 to 1800 in mathematics, physics, and chemistry.

This broad expanse of time encompasses the Protestant Reformation, the Thirty Years War, the French Revolution, and the American Revolution (though the Americas were not yet major players in the drama). The word "physics" began to show up in the English lex-

icon in the early 1600s—although meaning all of natural science[1]—
and the word "chemistry" about the same time—although described
as "a kinde of præstigious, covetous, cheating magick."[2] These two
endeavors were not necessarily regarded as distinct and both fell
under the umbrella of "natural philosophy" whose practitioners were
sometimes referred to as "virtuosos." The events of the Scientific
Revolution came in waves: The first revolution was in mathematics,
followed by a revolution in physics, which was followed by another in
chemistry. No sooner had the chemical wash receded than physics
and mathematics came crashing in again, followed by a tidal wave of
chemical breakthroughs.

The pattern is reasonable: The theory of chemistry builds with
the tools of physics, and physics builds with the tools of mathematics.
The discoveries of chemists often point the way for physical investigations and the needs of the physicists guide the efforts of the mathematicians. The three are bound in a complex symbiotic relationship.
But the interplay of the three takes time and such a broad timescale
can make the Scientific Revolution seem less revolutionary. Revolutions are supposed to be sudden and violent upheavals. Was the Scientific Revolution sudden? Was it violent?

The answer to the first question is yes. The period of time covered in the first part of this book was about 2,000 years, while the
time of the Scientific Revolution is about 300—so relatively speaking,
the Scientific Revolution was, to some degree, sudden. To the second
question, surprisingly, the answer is still yes. Most people don't think
of science as being an occupation that elicits violence, but in the case
of the European Scientific Revolution, people died—horribly. The
new age had its admirers, but it had detractors, too.

What were the objections of the detractors? Shouldn't scientific
progress have appealed to everyone? Today many of the objections to
change may be legitimate: Developments in science have brought
about pollution, ecological shifts, and societal stress. In Europe, in
the time of the Scientific Revolution, things could only get better.
Disease still caused massive plagues, surgery lacked anesthetic or anti-
septic, and industry relied on backbreaking manual labor. But the
objections raised to scientific progress during the Scientific Revolu-

tion were objections over a shift in power as much as a change in a way of life. The 1600s saw the rise of secular rulers, supported by science, who wrestled power from the Catholic Church.

The traditional Church, already under assault from the Reformation, fought back. So it was, in this enlightened age, that witchhunts hit their peak (the last witch burning in Europe occurred in 1782 in France)[3] and religious persecution was virulent. Those in science risked being seen as on the side of the secular and on the wrong side by the standards of the Church, a rather perilous place to be.[4]

But there was no safe haven. There was no social security. There was no welfare net. Tenure was rare rather than routine. When we speak of the infamous "publish or perish" now, we refer to the possibility of tardy promotion or early retirement—not being kicked out onto the streets to starve. In the second part of the Revolution, we will see less deprivation, but the memory of it will drive scientists to horde their secrets to impress their patrons, the secular rulers.

The problems that the secular rulers needed science to solve were manifold, including three of far-reaching importance. The first was to make the cannon a practical weapon. In the 1600s, the availability of gunpowder was still only a questionable advantage in war. True, cannons could hurl cannonballs at the enemy, but they had other machines that could do that as well. One invention, an uneven-arm throwing machine called the trebuchet, has been estimated to be able to fire a stone every twenty seconds.[5] These machines also ran on human power, not gunpowder, so they could be expected to work in the rain and were not subject to inopportune explosions. But plague had reduced the pool of available foot soldiers, and the cannon did not need food. So the utility of the cannon as a replacement for charging ranks became apparent. The problem was that the shooters didn't know how to aim. They knew that firing the cannon point-blank could certainly breach a wall, but fired horizontally the cannonball didn't go very far. To seriously damage a fortress, they had to roll the cannon practically up to the door, at which point the shooters could be picked off by archers. They also knew that pointing the cannon up at an angle increased the range, but then they weren't sure where the cannonball would land.

The second problem was naval battles. To ensure their coffers would grow, rulers needed to have uninterrupted sea trade and the ability to defend their ships. Cannons were more effective on the seas because a pointblank hit was more serious for a ship than a wall. So the problem became the ability to see the enemy coming. Enter the telescope. To understand how much of an improvement better telescopes brought, consider della Porta's method for dealing with capture by pirates: Take a potion to turn your urine red and rub your skin with an irritant to raise blisters. Why? In an age of pox and plague, these measures would at least keep the pirates from touching you, which is to say torturing you. With luck, the pirates would leave you adrift on your ship or allow you the opportunity to throw yourself to the mercies of the sea.[6]

The third problem was navigation. In addition to being able to see trouble coming, ship captains needed to know their location. On the featureless expanse of the sea it is possible to get blown off course and disoriented. The compass helped to locate north, but that was not enough. Pinpointing exact position required a longitude and latitude: longitude running from the North to South Pole and latitude circling the globe, at right angles to the longitude. Because Earth, tilted slightly, is traveling slowly around the sun in a yearly revolution, the latitude could be determined by the position of the sun relative to the horizon, which changed a bit every day, so the captains knew latitude as long as they knew the date. The Earth's spin, however, only takes twenty-four hours, so to know longitude they needed to know the position of some object, such as a fixed star, and a clock, and therein lay the rub. There were no clocks that worked well on shipboard. The problem was tackled by every scientist worth his or her salt. The British Admiralty offered a £20,000 prize for a method, but the prize went unclaimed until 1764 (and then the prizewinner got the glory but not the cash until he threatened to sue).[7] Out of the effort, however, would come some very interesting science.

These three problems, the need to maximize the range and aim of cannons, the need for better telescopes, and the need to determine longitude, encapsulate the problems facing the scientists of the Scientific Revolution. But why should a revolution be required to solve these problems? Because Aristotle's theories fell far short of what was needed. The only thing Aristotle had to offer on the subject of cannonballs was that heavy objects seek the Earth because it is their nature. This explanation met with a cold reception from the warlords of Italy. They needed to know how far a cannonball would fly before it sought its nature. In other words, they needed the philosophers to put a number to it. The Scientific Revolution was a turn from qualities to quantities. To be successful (that is, to receive support and ultimately a pension if they were lucky) the scientists realized they were going to have to put off the "whys" for a while and just focus on the "hows" and "how fars."

But knowing that you need information and being able to get it are two different things. In the Scientific Revolution, the problem was even broader: They first had to discover they *could* discover—that it was possible to know more than the Arabs and Aristotle knew—and that this information might be of some practical use. So it was that the Scientific Revolution began (following the precedent set in the Renaissance) with hungry artisans, mathematicians, and machinists finding they could better their lot by cleverness and talent and by rising above (or stepping around) the conventional wisdom. For those of a mathematical bent, the work was in the counting houses of the merchants. For those with mechanical abilities, the employment was in armaments and military devices. For those clever with chemicals, the path lay in medicine, gunpowder, poisons, or cosmetics. Cosmetics were important because women had so few employment options that physical attractiveness could be tantamount to survival. For the male mathematicians, physicists, and chemists the most comfortable positions were in the universities, if their reputations were good enough.

For the mathematicians, this need to establish reputations led to a lot of secrecy in hopes of elevating the demand for one's services. But reputation could also be established by publication—which sometimes led to publication of other people's secrets—which resulted in bad feelings. These streetwise mathematicians, with a life-or-death stake in their art, were the first to make concrete and useful discoveries that went beyond the findings of the ancients. The physicists who followed the groundbreaking mathematicians found it easier to wrestle with the long arm of Aristotle and to dispense with the search for ultimate truth in favor of the search for what ultimately works. These physicists, epitomized by Galileo, found the equations of motion that were needed for the field artillery and described the motion of the planets relative to the sun. The chemists, who had always felt more allied with magicians than mathematicians, gave up the search for gold in favor of the newly lucrative manufacture of medicines and gunpowder. Then the prototypical physical chemists, such as Robert Boyle, explored the properties of gas—the product of gunpowder reactions that causes the projectile to exit the barrel—and discovered mathematics applies to chemistry, too: a result that our medieval magician, Giambattista della Porta, would not easily have envisioned.

In the second wave of physics, Isaac Newton related force and motion, correlated the motion of cannonballs to the motion of the planets, and laid out the essential concepts needed to defend an atomic theory of matter. Along the way he found that he needed a new technique in mathematics, so he invented the calculus. When Newton tackled chemistry, however, he found himself embroiled in the mystic heritage of this art. Chemistry is a crafty beast and would continue to elude the saddle of mathematics, even when chased by Newton. It would take Antoine Lavoisier, an accountant and tax collector, to find the key to unlock the secrets of chemistry—the equations for the balance sheet of matter.

Chapter 6

Mathematics

The King was in his counting house, counting up his money; The Queen was in the parlor, eating bread and honey. The Maid was in the garden, hanging out the clothes; down came a blackbird and snapped off her nose.
English rhyme from the Renaissance period

Before beginning our journey into the Scientific Revolution, we need to look back at Renaissance mathematics. The Renaissance mathematicians set the stage for the Scientific Revolution by breaching the wall of the ancient Greeks. They solved equations that eluded even Euclid.

How were the mathematicians able to accomplish this feat? During these turbulent times, when scholars risked life and limb by expressing their ideas, what emboldened the mathematicians to tout their new thoughts? Were these mathematicians somehow more courageous and visionary then their kindred in the sciences? Hardly. Some of the mathematicians we will encounter will be some of our wiliest characters. But the mathematicians worked for the merchants and the merchants counted success in profits. As far as the merchants were concerned, the mathematicians could run roughshod over Euclid if their methods provided quicker or more accurate accounting. During these times alchemists/chemists were almost routinely hanged for their pursuit of what were considered the black arts. Proponents of the new celestial mechanics were hunted down and harried for what were considered blasphemous or anti-Aristotle utterings. But, as far as this author is aware, no mathematician was ever burned at the stake, and even those who had a close brush with such a death couldn't blame their mathematics.

The merchants also cared little for social pedigree, so mathematics provided an avenue to education and acceptance for the talented from any walk of life. No lives better exemplify the power of mathematics to bring a mathematician up a notch or two on the social scale than those of Girolamo Cardano (1501–1576) and Nicolò Tartaglia (1499–1557), two of the roughest of the ruffians ever to find a route to respectability through mathematics.[1]

It takes some understanding of Italy's history to appreciate Tartaglia's story. During the Renaissance, Italy was split into small defensive states by the rivalries between popes and powerful families. Born in 1499 in Brescia, a town situated in a fertile plain at the terminus of the Alpine pass, Nicolò Tartaglia lived in a perpetual battlefield. Tartaglia's father, a medieval mailman of sorts, died when Tartaglia was six years old and left the family in poverty. When a shift in political winds caused Brescia to be attacked by French troops with artillery, the twelve-year-old Nicolò tried to take shelter in the town cathedral. He nevertheless received several head wounds for his efforts, one of which tore his mouth and left him with a permanent speech impediment. For this, he gained the nickname of Tartaglia, based on the Italian "tartagliare," which means "to stammer." As an adult, he sardonically chose Tartaglia as a surname. His true surname remains unknown.

According to his own account, this tough little boy paid a tutor to teach him to read and write but ran out of money by the time they got to the letter "k." At that point, he resolved to teach himself and found his talent in mathematics. Soon he moved to Verona to become a tutor in mathematics. In those days, any mathematician with demonstrable knowledge and a pencil and paper could set up shop and attract paying students, so it became very important to retain one's reputation for being a skillful scholar. In the manner that magicians attract apprentices through keeping the secrets of their magic, these mathematicians attracted students to tutor by promising to teach mathematical secrets. To publicize their mathematical expertise, they would stage contests with other mathematicians. Often bets were made between combatants, so the contests could serve as supplemental income, but the main purpose was as an advertising venue to attract customers for their skills and students for tutoring. Though these contests began as a technique for increasing business, they became a tradition in European mathematics, even when the only stakes were reputation, but the battling was just as fierce, as we shall later see. It was in such an exhibition that the clash between Tartaglia and another major player in this story, Girolamo Cardano, took place.

A few words of explanation are needed here on an algebraic concept: the degree of an equation. The degree of an equation refers to the number of times the unknown is multiplied by itself. In other words, if you are looking for just the unknown, then it is an equation of the first degree. If you are solving for the square of the unknown (unknown times unknown), you have an equation of the second degree. The common names for these various types of equations hearken back to the geometric constructions of Euclid. An equation of the first degree is called a linear equation because a straight line can represent it. An equation of the second degree is called a quadratic equation because a quadrant is a square and the area of a square is found by multiplying side times side. An equation of the third degree is called a cubic equation because the volume of a cube is found by multiplying side times side times side.

By the 1500s methods for solving linear and quadratic equations were known. If a merchant knew that the vineyard produced two barrels a month, he could easily calculate how many barrels he would have in any given number months; this is a linear equation. In a similar fashion, another merchant might use a quadratic equation to determine the best loan. Let's say the merchant had a bad wine year and needed to borrow money. The merchant finds a financier willing to loan the money with a flat fee of 50 ducats per month, but the local loan shark offers what appears to be a cheaper, but increasing, rate: the age of the loan squared. The first week, the charge is one ducat: one times one. The second week the charge is four ducats: two times two. After the first month, four weeks, the charge is only up to 16 ducats: four times four. Of course, the loan shark's explanation will stop here because at this point the loan seems very reasonable. However, the merchant would be well advised to consult his or her mathematician because a little modern calculator math will show that in month three the charge is up to 144 ducats (12 times 12) and goes up to 576 ducats (24 times 24) in month six! The merchant can still take advantage of the initial cheap rates, however, if the merchant's mathematician properly projects when to pay off the loan—or when to leave Venice.

What does all this have to do with physical chemistry? A lot. There are several chemical quantities that depend on powers of factors greater than one. An immediately understandable example is the rate of reactions. The rate of a chemical reaction depends on the concentration of starting material, as any winemaker can tell you: The rate of fermentation depends on the concentration of sugar in the grapes. As can be seen by our example of a bad loan above, if the reaction rate increases as the second power of the concentration, this fact needs to be known to prevent potentially explosive consequences. So an understanding of the behavior of such mathematical systems had a direct impact on physical chemistry, but the effect was also indirect. Finding solutions to equations of the third degree— cubic equations—was the big breakthrough of the 1500s. Their solution demonstrated that knowledge existed beyond what was provided by the ancients.

There are several different types of cubic equations, and the Renaissance mathematicians treated them all as separate problems. For instance, there could be a cubic equation that included a squared term, a single power, or a constant thrown in for good measure. A rule for solving one of the simplest types of these equations—ones that contained just a cube term and a single power—was discovered by the mathematician Scipione Ferro early in the 1500s.[2] He did not publish his result but saved it as ammunition for the mathematical duels. He handed down his secret to his student, Antonio Maria Fiore,[3] and in 1535 Fiore challenged Tartaglia to a mathematical disputation armed with the following problem: A man sells a sapphire for 500 ducats. The original price he paid was the cube root of his profit. What was his profit and what was his cost? Tartaglia worked late into the night, found a method for solving the problem, presented the answer, and won the contest. Fiore, unable to answer Tartaglia's problems, lost. The victory put Tartaglia on top of the heap . . . for a while. Enter Girolamo Cardano.

Most of what we know of Girolamo Cardano comes from his own account. As an old man he opted to write an autobiography, some portions of which have been substantiated from other sources. It appears to be true that he was born in 1501 in Rome, the illegitimate son of an educated father (indeed, a friend of da Vinci) and a brutal mother. He was encouraged in his education by his father and earned a degree in medicine, but he could not be accepted by the College of Physicians of Milan because of his illegitimate birth. Obliged to practice medicine in the countryside, he took up mathematics as a hobby and as a way to help him supplement his income. Cardano was an inveterate gambler, as were many of the mathematicians of his day. They either came to mathematics in hopes of bettering their chances at games or came to gambling when they saw the power inherent in mathematics. So it was that Cardano was triply motivated in his pursuit of mathematics: He hoped it would improve his success in gambling, he hoped that it would add to his respectability, and he, too, realized, as many before him, that there were now profits to be made in the sale of books. He eventually published more than 200 works on everything from medicine to demonology, but it was while awaiting the publication of his first book on arithmetic, *Practical Arithmetic,* that Cardano learned that Nicolò Tartaglia knew a method for solving a cubic equation.[4]

Cardano approached Tartaglia and asked him for his solution, but Tartaglia, naturally, refused. Cardano accepted the initial refusal graciously, and continued to improve his fortunes in other ways. Cardano won acceptance to the College of Physicians, despite the circumstances of his birth, by impressive cures such as his treatment of the rather severe respiratory problems of the archbishop of Edinburgh. Cardano spent some time observing the archbishop's habits, noted that the condition worsened at night, and then ordered that the archbishop's feather comforter be replaced with a mattress of rags, which effected an immediate cure. Despite his success in medicine, Cardano continued to run up some rather substantial gambling debts, which necessitated a supplementary income.[5] To improve the value of his book on mathematics, he continued to ask Tartaglia for the latter's solution to the cubic equation.

Tartaglia continued to refuse until Cardano finally hit on the winning enticement: Tartaglia had been interested in artillery since the unfortunate incident of his youth and believed that he had conceived inventions that were good enough to sell. When Cardano promised to introduce him to the Milanese court for this purpose, Tartaglia consented to give Cardano his solution to the cubic equation, though Tartaglia did take some steps to protect himself. First he made Cardano swear not to divulge the secret, and second he gave the equation in the form of a poem to avoid having to write it down.

For the next several years, Cardano kept his oath, but then, perhaps inspired by Tartaglia's solution, he and his pupil, Lodovico Ferrari, worked on other forms of cubic equations and found solutions to them as well. Ferrari even managed to find a solution to a form of a fourth-degree equation. Seeing a way out of his oath when he heard that Tartaglia's solution was preexisting, Cardano and his pupil sought and found the notes of the original discovery. On the basis of this preceding discovery, Cardano published the solution to this particular cubic, too, but was careful to give Tartaglia credit along with the original discoverer.

As would be assumed, Tartaglia was furious. To be given credit is all fine and good, but the profits from the book went to Cardano. Threats and accusations flew until Tartaglia finally met Cardano's student, Ferrari, in a public mathematical challenge. Tartaglia was bested by Ferrari and had to leave in shame, and Ferrari, bolstered by his triumph, received offers of employment from royalty.

It is hard to establish in this cutthroat battle the villain versus victim, but none of the participants fared well in their future. Although he apparently did have a family at one point, Tartaglia died alone and in poverty. Ferrari died at the age of forty-three, probably poisoned by his sister in a family feud, and Cardano didn't do much better. He first endured the death of his son, who was executed for murdering his own wife, and then was arrested by the Inquisition for having the audacity to cast the horoscope of Christ. He was released after a few months in prison when he recanted and agreed to give up teaching. In his last book he wrote of his life and that, for all his misfortune, he had a grandson, a belief in God, and fifteen good teeth. But apparently he did commit suicide in the end. He had predicted the date of his death and did not want to be judged to be in error.

Yet it was Cardano and Tartaglia and the rest of the Renaissance mathematicians who showed that there was more to be known than the knowledge of the ancient Greeks and Arabs. But major hurdles remained before the full power of this new knowledge could be realized. The first is well demonstrated by looking at the beginning of the Cardano–Tartaglia–Ferro solution to a cubic equation as expressed by Cardano: "Cube one-third of the coefficient of the thing; add to it the square of one-half the constant of the equation; and take the square root of the whole. You will duplicate this and to one of the two you add one-half the number you have already squared and from the other you subtract one-half the same. . . ."[6] Renaissance algebra, following the Islamic tradition, was entirely rhetorical.

An algebraic textbook from the Renaissance might be unrecognized as such by anyone today. There were no neat equations set off by white space, no symbols for operators such as "+" for addition or "×" for multiplication, and no single-letter unknowns. All the relationships, manipulations, and operations were written out in words and the unknown was called *la cosa* or *the thing*. To subtract one from eleven, Fibonacci, the bookkeeper mathematician, would have written "a debit of one together with the sum of eleven becomes the number ten," and this statement would have been buried in a paragraph of text.

Moreover, proofs were worked out through geometry. To appreciate the cumbersome quality of a geometric proof consider the well-known distributive law of multiplication. If three girls and four boys buy shoes, they will buy fourteen shoes, whether it is calculated as two shoes per girl and two shoes per boy or the total number of children times two shoes each. For Euclid to prove this, he had to construct two rectangles, one of width two and length three for the girls and one of width two and length four for the boys. He then found the total area of the two rectangles together and compared it with a rectangle with a width of two and a length of seven.

Early in the fifteenth century some of the abacists (or professional mathematicians) began to substitute abbreviations for unknowns such as "C" for *cosa*, or "Ce" for *censo* or square, and "Cu" for *cubo* or cube. The resulting condensation helped, but things were still pretty awkward. Gradually the use of signs for "equals" and "square root" found their way into usage, promoted by mathematicians such as Michael Stifel, a cleric who used his mathematics for *wortrechnung* or word calculus: the interpretation of words through the numerical value of letters.[7] Using this system, he interpreted the Bible as predicting the end of the world on October 18, 1533. After his congregation waited in the church that whole day and nothing happened, they tossed him out. He eventually promised to give up predictions and was given another parish through the intervention of Martin Luther.

Another much needed step toward systemization was taken by François Viète (1540–1603), who turned the search for solutions to equations into a study of the structure of the equations themselves.[8] He was a lawyer, but his reputation for his mathematical skill was such that the king of France recruited him to the French court. One of Viète's tasks was to decipher intercepted coded messages, which could have intricate mathematical relationships as their key. He was so good at his "black art" (as mathematics was sometimes referred to by those in awe of it),[9] that some thought he should be denounced for sorcery. But the king of France was in such need of his services that he was willing to overlook, and encourage others to overlook, any hint of a sorcerer in his mathematician.

Viète was one of the first to substitute letters for all numbers in equations, which allowed a general, rather than specific, solution to be found. Until his time, each equation was solved as an individual problem. In other words, Viète's contribution is analogous to inventing the general concept of a bridge: When you come to a new river, instead of having to reinvent bridges, you just build your general bridge using dimensions specific for your river. This innovation compressed all the specific types of cubic equations that Cardano and Tartaglia struggled with into one general equation.

All these successes added to the arsenal of the mathematician, which further emboldened them. By 1585 Simon Stevin (1548–1620) wrote a book on decimal fractions that openly contradicted the wisdom of the ancient Greeks.[10] For us moderns, shopping for items costing dollars and cents hardly seems revolutionary. But for Stevin in these early times, proposing decimal fractions, the 0.98 of $24.98, took courage. The last time it had been proposed, the proposer, Hippasus, a member of the Pythagoreans, was supposedly pitched overboard.[11] Why the fuss? As Stevin had found, decimal fractions are easier to add, subtract, multiply, and so on, because they behaved just like whole numbers as long as you kept track of the decimal. The problem was that decimal fractions implied that numbers were continuously divisible—and Pythagoras had said that the number one was a pure ideal and that all numbers were built from this sacrosanct, indivisible number. Once the sanctity of one had been breached, what concept couldn't be thrown out? This idea threatened the underpinnings of mathematics as it was known. Many saw this as the beginning of the unraveling of Aristotelian tradition, and they were right.

In 1594 John Napier discovered that any number could be expressed as a power of ten, if you allowed for decimal powers: hence the birth of logarithms. The beauty of Napier's system is that numbers can be multiplied by adding their logarithms and divided by subtracting, which allows mathematical manipulations to be carried out easily, even mentally. As Napier observed, logarithms circumvented the need for long calculations that are a "tedious expenditure of time" and are subject to "slippery errors."[12] Napier was a Scottish Protestant, and his strong theological writings urging the king to "purge his house, family and court of all Papists, Atheists and Newtrals" probably saved him from persecution as a warlock by those who were suspicious of the origins of his mathematical prowess. Then, in the mid-1600s René Descartes through brilliant insight tied together the new algebra with the old geometry and laid out the path to the calculus.

By the time of Descartes we are foursquare into the Scientific Revolution with all its beauty, baroque elegance, and passion for order. Descartes, the scientist and philosopher, saw mathematics as a pure expression of logic and order.

Though known for independent wealth, good connections, and a penchant for sleeping late, Descartes lived as though he longed for the ruffian's life and looked for his intellectual roots in the lives of the back-alley mathematicians and low-level soldiers, but he didn't look too hard.[13] Descartes was a French Catholic at a time when Protestants and Catholics were at odds in France. He joined a Catholic army warring against the Protestants, but apparently did not participate directly in any battles. In 1622 he moved to Paris where he attended the theater and spent time gambling and fencing. Then, abruptly, he gave up his indolent life for that of a scholar. There are conflicting stories as to why he did this, but many seem to center on the well-recorded encounter between Descartes and an alchemist.

Apparently the alchemist, Chandoux (who shortly thereafter was hanged for counterfeiting),[14] proposed that there was a probability associated with experimental results; though you might not achieve precisely the same result with each measurement, this was not a reason for discounting the experiment. This belief would ultimately prevail— and, indeed, without it science could not progress—but Descartes, who longed for precision and mathematical method in science, argued that the result had to be precisely the same every time for it to be valid. On hearing the eloquence of Descartes in the public argument between Descartes and the alchemist, a radical Catholic approached Descartes and tried to enlist him in the Catholic cause. Descartes knew this was not the optimal time to involve himself in religious politics: Lucilio Vanini had been tortured and burned in 1619 for giving natural explanations of miracles, and the Parliament of Paris passed a decree in 1624 forbidding attacks on Aristotle on pain of death.[15] Descartes, who had a deeply rooted sense of self-preservation, found himself remembering his indiscreet youth and, with the crackling fires of the Inquisition in his ears, headed north to Holland.

It was in Holland that Descartes remembered some musings he had while keeping warm in a small heated hut near the battlefield and began to solidify his thoughts in writing. In 1633 Descartes was about to publish his book, which supported the notion that Earth revolves around the sun, when he heard that Galileo Galilei had just been hauled in front of the Inquisition for this belief. He demurred.

Fig. 6. René Descartes. Edgar Fahs Smith Collection, University of Pennsylvania Library.

In 1637, however, Descartes decided the tide had turned and published his *Discourse on Method*. In his method, Descartes proclaimed that all knowledge should be acquired through rigorous mathematical reasoning. The part that most interests us in this book is an extraordinarily long, 106-page footnote in which he provides an example of how the method might be applied to geometry.

In those days, calculations and proofs were done through geometry, which meant they had to construct geometric figures and measure lines. Descartes's insight was to recognize that geometric constructs, such as curves, could be represented by a set of points and therefore cast as equations. Many times equations can be manipulated more easily than the boxes and triangles of Euclid, and, conversely, some equations can be solved more easily when cast as curves.

As an example of how a curve might be cast as an equation and vice versa, consider the growth charts routinely used by pediatricians to track the health of children. These charts are based on an average growth rate that is vigorous when the child is young and gradually slows as the child matures. To determine if a child is growing normally the pediatrician generally rounds the age of the child, finds where the age line intersects the curve, and follows this point over to find the projected height. But the relationship between height and age could also be represented by an equation where the height would be equal to some curving function of time. Using this equation, the pediatrician could insert the exact age of the child and arrive at a more precisely predicted height. Before curves such as these could be realized, however, much work remained to be done. In particular, one important curve had to be solved first: the path that a particle takes when it acquires motion as a result of propelling force, such as a cannonball fired from a cannon, and under the influence of an attractive force, such as gravity. Understanding this curve would eventually lead to an understanding of the path of an electron under the influence of the force of a positive nucleus or the path of a particle feeling no force at all. Progress in this direction would be made by the Italian whose troubles with the Inquisition caused Descartes to defer publication: the mathematician Galileo Galilei.

Chapter 7

Physics

Still it moves.
Galileo Galilei, circa 1640

The description of motion is essential to physical chemistry. In the atomic model, atoms are always in motion. But Galileo's discovery of the fundamental characteristics of motion is only half the story. Galileo and his ilk represent what was revolutionary about the Scientific Revolution. They changed "I know" to "I don't know, but I can find out," and without this change we would not have been able to progress.

It is hard to give up the power of authority. Nobody likes to be wrong and nobody likes to admit they don't know. So the motivation for the ancients to speak authoritatively finds its origins in human nature. At the time of the European Scientific Revolution, the impetus was even stronger. As we have said, the world was a bit harsher during this period and employment was of paramount importance. For scientists to obtain and retain employment, they had to prove their worth and keep proving it. Nodding thoughtfully and saying, "Gee, that's an interesting problem but I really have no idea what to do about it" wouldn't pay the bills.

So when it came to the important concept of motion, most scientists had an explanation that they stated with as much authority as they could muster. Galileo had the advantage in that he had gained respect early in mathematics and with the design of scientific instruments. Once his reputation was established, he still couldn't bring himself to say, "Gee, I don't know . . . ," but he did manage a fairly forceful "You don't know, either." Then Galileo proceeded to find out.

Galileo was always outspoken, a trait he inherited from his father. Born in 1564 near Pisa, Galileo was the son of Vicenzo Galilei, a Renaissance rebel. One of those vital pieces of information imported into Europe from the ancient Greeks was the musical scale discovered by Pythagoras. Pythagoras found that if a stretched string makes a certain sound when plucked, then dividing the string in half and plucking it again will make the same sound, just an octave higher. Further, he found that when the ratio of lengths was 2:3, the two sounds weren't the same, but they blended: They harmonized. This harmonizing also occurred when the ratio was 3:4 and other whole number ratios. This is the principle behind playing the guitar: One moves one's fingers to hold the strings down at various lengths to make various pitches. While harmonic music is a mundane commodity in the modern world, imagine the ecstasy of Pythagoras when he discovered this simple fact. No wonder he believed in the purity and power of whole numbers.

Galileo's father was a proponent of a new way of tuning instruments that was not based on the rigorous whole number ratios of Pythagoras. He, as others around him (such as the multitalented Simon Stevin of decimal fractions and the Chinese prince Chi Tsai-Yü), decided it was more important to please their ears than Pythagoras. Watching and helping his father stretch and weigh strings of different lengths and thickness to obtain just the right sound, the young Galileo learned the value of experimentation.[1]

When he was seventeen, Galileo entered the University of Pisa with the objective of studying medicine, but soon turned to mathematics. He left the university when he was twenty-one, without a degree, but in those days reputation could take precedence over qualifications. He set up shop as a mathematics tutor in the tradition of Cardano and Tartaglia and eventually developed enough of a name to be hired as professor of mathematics at Pisa. Here he found that while mathematicians, in the service of merchants, were allowed to sidestep Euclid, professors had better respect Aristotle. Galileo, pointedly, did not have his contract renewed.

The point of departure between Galileo and the university fathers was that Galileo said flat out that Aristotle was wrong. Not maybe, not probably, but most definitively wrong—and he could prove it. The point of contention? Aristotle's conception of motion.

First let's reexamine the understanding of motion of that era as based on Aristotle's principles. Aristotle believed that motion was caused by an initial impetus and then perpetuated by air rushing in to fill the vacuum created by the object as it moved from its initial position. (Recall Aristotle's famous dictum: Nature abhors a vacuum.) If this had been all Aristotle said, there probably wouldn't have been a problem. But Aristotle, in his own effort to establish his authority, went one step further and made two predictions. The first was that the speed at which objects fall is proportional to their masses (that is, a hammer falls quickly while a feather drifts to the ground). Secondly, he said that in the absence of outside influence, all objects set in motion would eventually come to rest.

As we now understand (and as was demonstrated by an astronaut's experiment on the moon, which showed a feather and a hammer landing together),[2] air resistance is the factor that causes the feather to lag behind the hammer. But Galileo couldn't go to the moon, so it is said that Galileo debunked the first of Aristotle's predictions by dropping objects from the Leaning Tower of Pisa. Though the experiment most likely did occur, there is some question as to who conducted it. Some say it *may* be that Galileo did perform the actual experiment,[3] and some say Galileo hauled a cannonball and a musket ball up the steps himself.[4] But whether or not Galileo actually did the tower experiment, he performed the following thought experiment, which we can duplicate. Aristotle said that if the mass doubled, then the rate of fall would double. Galileo applied *reducio ad absurdum*: Could one envision a cannonball and musket ball being dropped together and the cannonball crashing to the ground, but the musket ball drifting down some several minutes later? However, it was an argument of qualities versus quantities at a time when quantitative science was just being born, so the university cited *any* difference in landing time as a proof of Aristotle's idea. The slight discrepancy caused by the resistance of the air was seen as a triumph for the old ways. The administration of the university of Pisa tossed Galileo out on the street, but he wasn't out for long. The same year he was appointed to the chair of mathematics at the University of Padua.[5]

As his appointments would indicate, Galileo was no slouch in the area of mathematics. He was so talented that he invented a calculating "compass," an early version of a slide rule, which provided some income from its manufacture and sale. His infatuation with gadgets would eventually lead to his articulation of the mathematics of motion, but gadgets would also be the reason for a delay of twenty years in the publication of his results. In 1609 he heard about the invention of a new type of gadget from Holland: a telescope.

Galileo made sure he was the first on his block to obtain one from a Dutch lens maker.[6] The first telescope he obtained was about as powerful as modern binoculars, but this was impressive enough. When he demonstrated the device in Venice, the university doubled his salary and he received tenure for life. By December of the same year, Galileo had designed his own telescope and improved it to the point that he could see mountains and craters on the moon. He also discovered the four largest moons of Jupiter, the phases of which he proposed might make a nice clock for determining longitude. He published these findings in *The Starry Messenger*, and the book's fame won him appointment as court mathematician at Florence. This position released him from his teaching duties and allowed him to do research full time. It should be noted here that he was offered the title of "Philosopher," which carried great prestige in itself, but he insisted in retaining "Mathematician," too. He saw science as a mathematical enterprise and wanted that reflected in his job title. The civic authorities obliged him because his improvements to the telescope enabled them to see ships coming sooner, but the religious authorities began to realize that Galileo was seeing things they'd rather he didn't see—and he was ready to tell the world. As can be discerned from his difficulties with the university fathers of Pisa, Galileo never was a conformist nor a soft-spoken man. He even got into trouble because he refused to wear the regulation academic regalia at all times. He wrote a poem in which, among other things, he complained that you couldn't go into a brothel in regalia.[7]

The new recognition he received emboldened him further. For instance, he began to back the theory of Copernicus. Back in 1543, a quiet cleric by the name of Nicolaus Copernicus had published a book entitled *Revolutions of the Heavenly Bodies* in which he proposed that the planets moved around the sun rather than the other way around.[8] He was very insightful but he wisely waited until he was on his deathbed to allow the publication of his book. The publisher, in an equally prudent act of self-protection, added an editorial preface in which he stated that Copernicus was just offering a mathematical model that made astronomical calculations, such as calculating the date of Easter, easier, which it did. However, there is little doubt that Copernicus believed what he wrote. He envisioned the orbits of the planets as circular, which they are not, but writing his book put a name to what scholars were gravitating toward anyway. The "Copernican system" became a rallying cry for the Scientific Revolution.

When Galileo backed the Copernican system, as did many others, his investigations did not immediately bring him to loggerheads with the Church. There is every indication that Galileo wished to be a good Christian and had no quarrel with the religious establishment. True, there had been that nasty burning of the magician Giordano Bruno in 1600, supposedly for the same belief, but then that international wanderer had rather recklessly offended established authorities in other ways as well.[9] Galileo probably would have hesitated if he thought he would cause discomfiture in the Church, but in fact, he discussed his ideas openly, thinking he was having intellectual discourse. The popes were then as they are now: educated, intelligent men, knowledgeable in the sciences as well as the arts.

In 1616 Galileo received a gentle scolding from the pope and a cardinal for too publicly and too confidently supporting the Copernican system. They said that this matter was best left to the Church fathers and Galileo obliged. However, when a new pope, Urban VIII, was elected in 1623, Galileo sensed a kindred spirit. He visited Rome

Fig. 7. Galileo Galilei. *Symbols, Signs & Signets,* © 1950 by Ernst Lehner, Dover Publications, Inc.

to show Urban the telescope and in 1632 sent him a copy of his new book, *Dialogue on the Two Chief Systems of the World,* in which he openly embraced Copernicus's theories.

Unfortunately, the book arrived at a bad time. It had to be taken apart and fumigated before it could be handed to the pope (the plague still raged in waves) and when Urban received it, he was deeply embroiled in political problems. The Thirty Years War, which started when some Protestants threw some Catholics out a window,[10] was threatening to tear the Church apart (and in fact it did), so it can be understood why the pope decided that the sun and Galileo would have to wait until he had time to deal with it. Consequently, the book was reviewed by a cardinal who was critical of the pope's handling of the war and of Galileo, so the pope, in an act of appeasement, agreed to crack down on Galileo.[11] Eight months after the book came out, Galileo was invited to appear before the Inquisition.

Galileo was, by all indications, shocked. He had taken pains to go through all the proper channels and receive all the proper approvals for the book that were required. He wrote back that he was ill, which he was, as certainly anyone would be in such a situation. Finally in June, Galileo could delay no longer and arrived in Rome. Told that he could be interrogated as to true intent, and shown the instruments of torture, Galileo wisely recanted. In what was perhaps an act of kindness, he was merely put under house arrest—for life. Despite the indignation, illness, and, in the end blindness, Galileo used the enforced solitude to finally record the observations on motion he had made in his youth. His last book, *Discourses and Mathematical Demonstrations Concerning Two New Sciences,* was sent to the Netherlands, where the Inquisitors had no power, and published in 1638.

Among other topics in *Discourses* was Galileo's mathematical analysis of motion, including his assertion that all objects in free fall, in the absence of air resistance, fall at the same rate. Galileo's discovery can be easily demonstrated by dropping a set of keys from significant height (10 meters or greater) and timing how long it takes them to hit the ground with a stopwatch, a device Galileo didn't have, but would have found wonderful. Keys work especially well because they are small and heavy and make a loud noise when they hit the ground.

Switching to more keys or fewer keys or even attaching a bowling ball to the keys doesn't change the time it takes to hit the ground. In addition, through careful analysis Galileo found that the velocity for an object in free fall increases, and this increase is also the same for all objects in free fall: the gravitational acceleration of 9.8 meters per second squared, everyone's mantra from Physics 101.

Another bit of brilliance from Galileo was the concept of inertia: the idea that, in the absence of outside forces, a body in motion will stay in motion and a body at rest will stay at rest, directly denying the dictum of Aristotle. It is fairly easy to see how one might deduce that a body at rest stays as rest, but how does one prove that an object in motion would stay in motion? The answer can be found in Galileo's clever experimental design, careful measurements, and the use of an extension of inductive proof called extrapolation. The first thing he did was to select massive metal balls to minimize the influence of air resistance, then polish them to minimize the effect of friction. Galileo found his carefully polished balls rolled more and more slowly when he released them from the top of gradually flatter inclines, but they did not come to rest at the bottom. He mentally extended, or extrapolated, his results to a plane with no incline whatsoever and said that, in the absence of friction, a ball set in motion on this surface would continue in motion forever.

It is inertia that necessitates seatbelts in cars: A body in motion in a car will stay in motion, even after the car stops, unless a force is applied, such as a seatbelt. The concept of inertia is also very important to the analysis of molecular motion. Molecules of air do not experience air resistance—they *are* air resistance—so once set in motion, they stay in motion. From this fact many physical properties of matter, such as pressure, would soon be derived.

So Galileo's work had many direct ties to physical chemistry— and one curious indirect tie, too. Having found love with one not of his social strata, the beautiful Marina Gamba, he had several illegitimate children, one of whom was particularly dear to him, a daughter. Because of her illegitimate birth she was considered unmarriageable, and remained cloistered as a nun all her life. Her celestial choice of name on receiving her habit, Maria Celeste, shows she shared her father's fascination with the skies, and she remained her father's closest confidante through correspondence and by careful solicitous care for his health. A moving account of Galileo's life, *Galileo's Daughter*, by Dava Sobel,[12] tells the story of how Galileo's disciple attempted to have his master's body properly entombed but was prevented from doing so by a still sensitive political situation. Because of this Galileo was consigned to a virtually unmarked grave until 1737 when his remains were moved. Beside Galileo they found the body of his daughter, who had preceded him in death. She was carefully moved with him.

Why does this daughter represent another of Galileo's connections to physical chemistry? Galileo's daughter was a chemist.

Chapter 8

Chemistry

. . . we began to pour quicksilver into the longer leg . . . until the air in the shorter leg was . . . reduced to take up by half the space it possessed before . . .; we cast our eyes on the longer leg . . . and we observed, not without delight . . . that the quicksilver in that longer tube was 29 inches higher than the other.

Robert Boyle, circa 1660

Chemistry has roots in the healing arts and physical chemistry shares those roots through the measurements of the medicine makers. Galileo's daughter, Maria Celeste, is part of this heritage because she was the apothecary at the convent where she lived.

A misunderstanding that has come down to us in modern times is that the early remedies for illness were just magic and superstition. Some were. But some worked. Purgatives were important in northern climates where fruit was difficult to keep through the winter. Some herbal brews were sources of vitamins. Other remedies existed for pain and some were antiseptics. But one area where the medicines were a colossal failure was preventing or curing the plague. The first symptom of the plague was swollen lymph nodes under the arm or in the groin. These were the buboes that gave the disease the name bubonic plague. Most treatments centered on the buboes and ranged from burning or lancing to leeches or cauterizing with caustics. The dressings applied to the wounds ranged from animal dung to poison ivy to crushed glass.[1] In most cases, however, death occurred in a week, if not sooner. People were as ignorant of the causes as they were of the cures, but some realized that contact with diseased people could spread the disease and that cleanliness and fumigation were guards against the disease. So Maria Celeste is to be congratulated: The plague did not breach the walls of her convent during her tenure as apothecary.

There are wonderful insights into convent life contained in Maria Celeste's letters to her father. His replies to her are lost, perhaps destroyed after her death for fear of association with an Inquisition convict. But from her letters we learn that Galileo was bedridden several times with various maladies that she attempted to ease by sending him remedies from her stores. But she did not write of her cures and concoctions except to mention successful cures occasionally in letters, such as curing her father's messenger of face lesions. Fortunately others who experimented with substances wrote of their chemical knowledge. Of these, Marie Meurdrac is an outstanding and interesting example.

Not much is known of Marie Meurdrac's life, which may have been to her advantage while she was still alive.[2] Consider the case of another woman interested in chemistry, the unfortunate metallurgist, Martine de Bertereau du Châtelet, who toured the mines of France with her husband and urged the king to further exploit his mineral resources. She wrote of the science of mining and the art of assaying and smelting ores, but her advice must have fallen on some unreceptive ears because she was imprisoned for witchcraft. She died in jail before she could come to trial.[3] In contrast to the self-promoting du Châtelet, Meurdrac kept a much lower profile: She painted herself as an unpretentious person who had some knowledge that she hoped to be of value. What we know or can glean of Meurdrac comes from her book, *The Chemistry, Benevolent and Easy, Favorable to Women*.[4] This book, first published in 1666 and approved by the Regents of the Faculty of Medicine of Paris, went through several editions and was translated from the original French into German and Italian. In it, Meurdrac refrains from making claims of miraculous cures such as were expressed by della Porta or Paracelsus, another chemical physician of the same era and one whose work is more commonly cited in more comprehensive histories of chemistry.[5] Instead she says simply that she has tried several of her recipes many times and has found them to be effective. We will examine Meurdrac's book here in lieu of similar books such as those by Paracelsus for two reasons. The first is that Meurdrac's book is well organized in contrast to the rambling style of Paracelsus, and the second is that Meurdrac represents a new direction, one that will be of pivotal importance to the Scientific Revolution: She emphasizes the importance of carefully consistent measurement.

In the absence of a uniform system of weights and measures (we must wait for Lavoisier and his colleagues to devise the metric system) Meurdrac relates weights to standards such as common coins. For smaller weights, she has more of a problem. The basic unit of mass she uses is a grain, which she defines as the weight of a big kernel of barley. For a measure of volume she uses "a pinch," defined by what three fingers can pick up. Meurdrac also gives detailed descriptions of her chemical operations.

In the first part of her book she discusses distillation, sublimation, filtration, digestion, putrefaction, fermentation, and precipitation. She provides detailed descriptions of different kinds of fires and furnaces as well as various vessels and how to seal them. It is obvious that she performed all these operations herself because she speaks of them knowingly, such as her discussion of the tarry material that is inevitably left in the bottom pot in organic distillations. This usually black, gooey residue of poorly defined carbonaceous material has been called GOK (God only knows) but Meurdrac prefers to call it "feces." She briefly discusses the current theory of chemistry, which varied from authority to authority, but Meurdrac settled on the one also put forth by Paracelsus: Salt, sulfur, and mercury are the three building blocks of all matter. She does not dwell on any particular proof or discussion of ramifications of this theory, but merely states it. Apparently she did not find it too helpful.

The second part of the book tells how to extract and concentrate the active component of medicinal herbs. She gives methods for making salts, tinctures, oils, and essences. She is careful to caution against using heat in extractions when not necessary, which leads one to believe she had indeed experimented with her remedies: Heat can destroy an active ingredient and ruin the efficacy. She then goes on to describe the application of the medicines as well as the cosmetics that can be made from vegetable, animal, and mineral materials.

Meurdrac's book is much narrower in scope than della Porta's eclectic collection of magic. There is no discussion of magnets or gunpowder because they do not relate directly to Meurdrac's medicines and cosmetics, and Meurdrac offers numerous medicines for women and children, which della Porta does not. Specifically, she recommends medicines for easing the discomfort of childbirth and treatments for worms in children: not a trivial problem in Europe in those times. Unfortunately, treatments that attacked the worms also tended to attack the host. But Meurdrac assures her audience that her remedies will work without undue discomfort and even offers the suggestion that the children's stools be inspected in subsequent days for evidence of dead worms. Some of the medicines she prescribes have recognizable efficacy, such as calcined eggshells for stomach upset, which contain calcium carbonate, a substance found today in over-the-counter antacids. She also recommends ointments for skin eruptions that contain either antimony or mercury salts, which are known to be effective antibacterial agents.

Meurdrac's work also contains her share of medicines whose reputation is based more on hearsay than experience or, if based on experience, the experience was coincidental: There were some people who contracted the plague and survived and certainly many documented cases of doctors and healers who worked among plague sufferers and lived to tell their tale.[6] So it was a natural temptation to look back and try to reconstruct whatever the survivor did in hopes of repeating the cure, which inspired the more creative treatments. During the era of the mid-1600s, preparations from the human body were prized as medicines.[7] Meurdrac speaks of tonics made from extracts of human blood, skull, and marrow, though no mention is made of how the starting materials were procured. A bit of reflection on the current elaborate measures for handling and disposing of medicinal waste shows how reckless a practice this was in an age of plague. Such considerations lead one to suspect that the rats weren't the only agents in the diffusion of the disease.

But for all the imprecision and hearsay, the book's emphasis on repeated, tested recipes and on weights and measures represents a very important step forward. Careful measurement would be as important to the evolution of chemistry as it was to physics, but chemistry would still have a much more difficult time throwing off the shackles of Aristotle than physics or mathematics. The reasons are manifold, but two can be singled out here. The first was that it was very difficult to get reproducible results in chemistry in those days, even if one allowed for experimental error. Methods for analyzing for purity were rare, and materials can react very differently if they contain even small amounts of contaminants. For instance, if one tries to make bread with self-rising flour instead of regular flour, the added salt and carbonates can inhibit the action of the yeast. A baker might wonder if demons and elves were making mischief if the same recipe were followed and the bread rose on one day and went flat on the next. The second reason is that the notion of qualities was very useful in chemistry. While the physicists were searching for quantities, the chemists were comfortable with Aristotlelian qualities. For chemists, it was helpful to know what materials had acidic qualities and which were explosive. Breakthroughs emerged when scientists who practiced both physics and chemistry found they were able to note qualities and put numbers to things, too. In this respect, one of the most notable men of this time was Robert Boyle.

Robert Boyle was born to a landed, wealthy family in Ireland in 1627, the youngest of fourteen children.[8] He was educated at home, then in England, and then on continental Europe. His family brought him home during the Anglo–Irish War and English Civil

Fig. 8. Robert Boyle. Edgar Fahs Smith Collection, University of Pennsylvania Library.

War, but when the smoke cleared, Boyle, having been disappointed in his marriage plans, went to live in his sister's house in London. His brother-in-law was reportedly a brutal drunk, so his sister no doubt welcomed his presence, and for whatever reason he made himself a permanent part of the household, to the extent of setting up a laboratory in the house and making himself comfortable as a self-supported scholar. His fascination with chemistry is said to have arisen from an interest in medicines. A sickly man, he probably suffered from kidney stones. But for those of us who practice the art of chemistry, we know that there is a fascination with the bumping and boiling of chemistry that engages and entraps those with no motive at all. So while Boyle may have started out with an interest in developing medicines, he soon wandered far afield.

Boyle investigated the qualities of various materials and attempted to classify some types of chemical reactions. For instance, both acids and bases can be classified as corrosive materials, but they are, in a sense, opposites. An acid will react with a base to produce water. This type of acid–base reaction is called neutralization and is the reaction that is responsible for the efficacy of calcium carbonate in treating indigestion. If the stomach is producing excess acid, the calcium carbonate, a base, reacts with the acid to produce innocuous water. One of the experiments Boyle employed to identify acids and bases can be reproduced in any kitchen with purple cabbage, though Boyle may have used colored flower petals or other brightly colored vegetation. If purple cabbage is finely chopped and cooked a bit in water, it makes a purple liquid that can function as an acid or base indicator. Cabbage juice turns vinegar, an acid, crimson red. Cabbage juice turns baking-soda water, a base, royal blue. It is also interesting to test clear carbonated water or a clear soda drink. These drinks test acidic. But though Boyle was enchanted by his acids and color changes, it was while working with air that he confirmed and studied the relationship that brought mathematics to chemistry.

Boyle and his cronies formed a group known as the College for the Promoting of Physico-Mathematical Experimental Learning, which gained the king's approval in 1662 and became The Royal Society for the Improvement of Natural Knowledge. The Royal

Society still exists today and is a major force in the proliferation and propagation of science—but back then, things were a bit different. Anyone who could pay dues could join. The king lent his name but not his money.

Much of the experimentation was just curious poking and prodding with no systematic approach, which included animal experiments that were quite cruel, though there appeared to be little cognizance of the cruelty. These experiments included live dissection and the employment of the newly invented air pump to evacuate the air from a bell jar in which some creature was ensconced. They did experiments on all manner of beasts and insects and there is even a note in Boyle's own hand that they thought to experiment on a fetus but could not procure one.[9]

Boyle's interest in air pumps and the effect of air led him in 1661 to explore the relationship named after him, Boyle's law. Although Boyle did not claim credit for discovering this relationship—he had heard of it from others who had tried it—he did study the effect and disseminate the knowledge. The law states that the volume of a gas is inversely proportional to the pressure: If pressure is applied to a gas in a confined space, the volume decreases (if you squeeze on a balloon it shrinks) and the amount it shrinks is directly proportional to the pressure applied. If you squeeze twice as hard, the balloon shrinks twice as much. Boyle demonstrated this effect by trapping a bubble of air in a sealed side of a U-shaped tube with a plug of mercury. He then poured mercury into the side of the "U" opposite the bubble and watched the bubble compress. He found that when the height of the column of mercury increased by a factor of two, the bubble compressed by a factor of two. This discovery delighted Boyle. They had discovered a mathematical relationship as predictable as the acceleration of Galileo's cannonballs in free fall.

Boyle's experimentation was indeed seminal in the development of physical chemistry for two reasons. First, gas models would become extremely important in establishing the behavior of matter and the atomic theory. The gas phase makes for a good, simple model with many of the confusing and confounding factors, such as the stickiness of particles, removed. At sufficiently low pressure and high temperature, gas particles behave as though they are alone in the universe. They are still as sticky as ever, but they are not close enough together to have it matter. At room temperature and at normal atmospheric pressure, the ratio of the volume occupied by one nitrogen molecule to the surrounding empty volume is equivalent to a population density of about one person every 200 square feet. In addition to being far apart, gas-phase particles whiz around at a fairly good clip. Just as it is more difficult to catch a fast ball than a slow ball, the importance of the stickiness factor decreases as the relative speed increases.

The second reason Boyle's experiment was so important was that this experiment showed that yes, mathematics applies to chemistry, too—at a time when this possibility was very much in doubt. People believed in magic, and most chemistry seemed mystical. Boyle was very credulous himself and listened to and followed all tales of strange events and transmutations. His credulousness is really not so difficult to understand. The physics of chemical reactions would remain shrouded in mystery for a couple of hundred years more. When the basics of reactivity were finally understood to some extent, they required a physics and mathematics that Boyle could not even imagine.

Boyle was not alone in his bewilderment. Our next hero would make tremendous progress with his major insights and innovations in physics and mathematics—but he, too, was mystified by chemistry. Sir Isaac Newton, it seems, was confounded by alchemy.

Chapter 9

Mathematics
and
Physics

> *I thereby compared the force requisite to keep the moon in her orb with the gravitational force at the surface of the earth and found them to agree pretty nearly.*
>
> Isaac Newton, circa 1690

Sir Isaac Newton an alchemist? The inventor of the calculus, the codifier of physics, and the lighthouse of logic in the Scientific Revolution? A serious student of allegory, mysticism, and magic? Yes. Virtually all biographers of Newton recognize the fact that he indulged in alchemy,[1] but most treat it as an unfortunate eccentricity and relegate it to the end of the story. They point to Newton's other occasionally neurotic behavior or the common practice of self-prescription in those days (his mentor, Isaac Barrow, died from a drug overdose)[2] and say that Newton's interest in alchemy may have come from a drug-induced delusionary state or perhaps mercury poisoning.[3] Even those who see it as a significant part of his research may explain it as a religious attempt to unite spirituality with science.[4]

111

While this is certainly a plausible explanation given his deep religious convictions and his religious writings, the actual explanation may be simpler than that. Newton may have delved into the illogic of alchemy because it was the logical thing to do.

Consider, if you will, the reaction of the human race if an alien spacecraft should arrive on Earth and in it were found, among other things, the cure for all forms of cancer. Everything else in the spaceship would immediately look superior, too. This analogy works fairly well for the arrival in Europe of the knowledge of the ancient Greeks and Arabs. The geometry of Euclid had great practical power and the Renaissance mathematicians built on this foundation. The medical tradition of Galen included many useful remedies and the rediscovery of the harmonic tones described by Pythagoras must have felt like stumbling on a key to the cosmic order. So after Newton invented a reflecting telescope, broke white light into its component colors (creating a non-mythological explanation for rainbows), and found mathematical models for gravity, he turned his attention to the mystery of chemistry and looked to the ancients as the logical place to start. After all, the calculus found its origins in the ancient works of Euclid. Newton's success in physics came from studying the works of Ptolemy, Copernicus, and Kepler. It must have seemed reasonable to him to delve into the literature of alchemy to find the secret of chemistry.

Why was he not put off by the alchemists' claims of mystical forces? Probably because he proposed the mysterious force of gravity that acted through the vacuum of space. Why wasn't he bothered by the alchemists' obscure language and symbolism? Probably because he occasionally hid his own mathematical tricks in anagrams and used other devices of secrecy familiar to the wily Renaissance mathematicians. He, too, had been of humble origins and knew the value of keeping a trade secret.

Newton was born on Christmas day the same year Galileo died, the son of an illiterate English farmer.[5] He was a sickly child, severely undersized, not expected to live. His father died before he was born and his mother then married a man who did not want to take on a ready-made family. His grandmother raised Newton until he was eleven, when his mother was widowed again. He had an elementary education that his mother did not intend to continue because she thought he would be a farmer like his father. However, a schoolmaster recognized his ability in mathematics, and his mother recognized his inability in the field, so they decided to continue his education. He was able to pay the fees at Trinity College when he was nineteen and impressed enough people to obtain a scholarship three years later, a fellowship three years after that, and a professorship two years after that— a stellar rise in anyone's book. He was not a successful professor, and as attendance at his lectures was optional, he sometimes read to an empty room to fill the terms of his contract. How then had he received his professorship? Mainly by the recommendation of his mentor and the person who vacated the position in his favor: Isaac Barrow. What so impressed Barrow? Newton's invention of the calculus.

It is proper to use the article "the" before "the calculus" for the same reason you would say the British Isles, not just British Isles. The calculus is a system of calculating that encompasses two specific operations: integration and differentiation. These two mathematical operations are the inverse of each other as multiplication and division are inverse operations, just a bit more difficult to describe. What does integration do? It finds the area bounded by a curve. To understand how this works, imagine estimating the area of a curved, landscaped lawn that is bordered on one side by a sidewalk. You could accomplish this by walking down the sidewalk and laying down planks until you had covered the whole lawn, trimming the planks to shorter lengths when necessary or splicing on more length if needed. You could then measure the length of each board and multiply it by the width to arrive at an approximation to the area of the lawn. The measure would not be exact because the planks are rectangles and the lawn is a curve, so there would be an overshoot or undershoot with each board. In the calculus, the planks are made vanishingly thin, so the approximation to the area is very good indeed.

What does differentiation do? It finds the slope, or rate of change, of a curve at any point on the curve. The grade of a road is a measure of its slope. If you are on a road with a forty-five degree grade, then for every two feet you travel in the horizontal direction you also travel two feet up. This slope is pretty steep and you would need a powerful car to pull any kind of load up a grade like this. Conversely, a downward grade of forty degrees would require good brakes and good driving skills. The role of differentiation is to find the grade of the road, that is to say, the slope of the curve. What Newton managed, building on the work of others, including Barrow, was to come up with a general method for finding the slope of any curve at any point. But more than that, he was able to show that integration and differentiation are inverse operations.

To imagine why this is so, picture an accountant whose job is to keep track of construction costs. The accountant finds that the total cost of fencing material on one job is $100 the first day, $200 the next, and $300 dollars the next. By looking at the steady increase in the total cost of the fence, the accountant, without leaving the office, knows that it is a nice, level fence because every day's work adds the same amount to the total. But on another job, the total cost is $100 the first day, then $300, then $400, then $600—changing by $100, then $200, then back to $100, then back to $200. Again, without leaving the office, the accountant can surmise that this fence has an up-and-down shaped top because of the way the total cost changes each day. By knowing the rate of change (derivative) of the total cost (the integral) per day, the shape of the curve describing the fence can be ascertained.

But Newton had been a failure as a farmer and the shapes of fences didn't interest him. What did interest him was the shapes traced by the orbits of the planets. Newton used his calculus to show how the motion of the moon was related to the motion of the planets and the motion of objects in free fall on Earth. In the process he invoked an enigmatic force called gravity.

Here is the legacy that came down to Newton. After Copernicus's statement of the heliocentric theory and the resulting furor, the story passes on to Tycho Brahe, a quarrelsome man who had his nose cut off in a duel and had to wear a false nose all his life. Brahe was given an island by the Danish government to set up an observatory for gathering astronomical data for navigational purposes, but he also built a lab and did alchemical experiments as well. His tenants hated him so much they burned his home, observatory, and laboratory when he left. He was, however, an excellent astronomical

Fig. 9. Sir Issac Newton. Edgar Fahs Smith Collection, University of Pennsylvania Library.

observer and kept meticulous records. After his death, his assistant, Johannes Kepler, obtained the records and determined to find a pattern in all the numbers. Kepler believed strongly that the ancient Greeks had a special insight, and because they believed there would always be simple whole-number harmonies, Kepler searched for whole-number relationships in the orbits of the planets. He found first, by careful analysis, that the planets traveled in elliptical, not circular, orbits (you obtain a circle if you cut an olive through the middle, an ellipse if you cut it end to end), then spent years searching for a simple relationship between the length of the solar orbit for the planets and their distance from the sun. After trying many relationships that did not work, Kepler finally found one that did: The square of the time of rotation divided by the cube of the distance from the sun is a constant for all the planets. He published his findings in a book, *Harmonies of the World*, in 1619. Newton used his calculus to show that the force that caused such heavenly orbits also predicted that objects on Earth should fall at a constant acceleration.

Newton did not immediately publish his findings. In fact, it took about twenty years before his friend Edmond Halley convinced him to write them up and submit them for publication by the Royal Society of London (the same that fostered Boyle). The Royal Society, however, did not want to be involved and Halley ended up underwriting the publication of Newton's book, *Principia*, himself. The reason? In the twenty years that Newton had kept his secrets others had come up with similar ideas, and in the tradition of the Renaissance mathematicians, a fight would ensue over who discovered what.

Isaac Barrow, Newton's mentor, had also been looking for the secret to the calculus and was duly impressed by Newton for finding it. There were many other mathematicians working on the problem of the calculus, too, such as the Japanese mathematician Seki Kowa and a German, Gottfried Leibniz. It is undoubtedly true that Newton preceded Leibniz in his discoveries, but it is also true that Leibniz published before Newton.

The German Gottfried Wilhelm von Leibniz was an interesting sort of entrepreneur who traveled all over Europe selling his services as a person who could research lineage and for the right price find evidence of royal heritage—or for a better price, title to family lands or throne. He took up mathematics as a hobby, no doubt to pass time in long carriage rides from court to court and perhaps as a way of impressing learned clients. Leibniz had sufficient talent to see the relationship that Newton had seen ten years earlier but that Newton had deemed to keep to himself. Leibniz's article in 1684 in *Acta Eruditorum* introduced methods for finding the derivatives (slopes) of diverse curves. He called the article "A New Method for Maxima and Minima as Well as Tangents, Which Is Impeded neither by Fractional nor by Irrational Quantities, and a Remarkable Type of Calculus for Them."[6] He published a second article, "On a Deeply Hidden Geometry," in which he explained integration, a few years later. When Leibniz published his discoveries, Newton was angry and defensive, claiming that Leibniz must have stolen the idea from him. Because of the nationalistic nature of the dispute, contests inevitably arose. In the tried-and-true fashion of the backstreet Bologna mathematicians, but in correspondence rather than face to face, Leibniz threw problems at Newton and Newton responded with solutions, but the debate was never resolved.

What was the reason for Newton's reticence to publish that allowed him to be scooped by the continental mathematician? Some have said it was because Newton was a naturally quiet man (at one point he served in Parliament and his only recorded words were a request to open the window), and some have said it was because Newton was sensitive to criticism. Accordingly, he did not want to invite too close a scrutiny because he felt he had obtained his results without rigorous proof, which he knew would be criticized. But the reason may be simpler: Newton had advanced himself from a farm to London and had been appointed Lucasian professor, a position now held by Stephen Hawking, on the strength of his abilities in mathematics. No wonder he was close mouthed. He didn't want to give away the secrets upon which his reputation was based. In addition, he had an application of his mathematics in mind that he knew would guarantee the prominence of his position: the analysis of motion.

In modern parlance, Newton explained that the force produced by a moving object is equal to its acceleration multiplied by its mass. This concept can be explained by picturing the all-too-common experience of a car crash. To begin with, consider acceleration: Acceleration is a change in velocity. Stomping on the accelerator pedal in a car causes a change in velocity. Acceleration is felt, but constant velocity is not. When you are moving at constant velocity you don't feel the tug of acceleration: You can get up and move around in an airplane traveling at constant velocity and it can seem as though the plane were still sitting on the ground. But when the plane accelerates, takes off, you feel a definite tug. This tug is the result of acceleration. When an automobile is in a crash, the velocity also changes: from the speed you were going to zero. The severity of the damage will depend on two factors: how fast the cars were going when they hit and whether the crash was with a truck or a compact car. In other words, the force depends on the acceleration and the mass. As we will shortly see, one of our best pieces of evidence for the atomic model is that atomic-sized air particles—constantly whizzing around and experiencing mini car crashes with objects such as the side of a balloon—create a force on the area of the balloon that we measure as air pressure.

Newton also used his calculus to show that the force causing the balls to roll down Galileo's wedge was the same force that held the planets in their orbits: gravity. His results brought him the criticism he dreaded, if indeed this is what he was dreading. The idea of a force acting through the vast emptiness of space was a difficult pill to swallow. It was too magical. Too mystical. Too alchemical. But Newton made no attempt to explain the force. He just pointed out it was there.

Newton's identification of gravity as a force acting at a distance, a force acting without direct contact with its target, will be extremely important to future theories in physical chemistry. Electrons in atoms are described as under the influence of an attractive force acting at a distance as gravity does. Intermolecular attractions between atoms and molecules are forces acting at a distance. In their search for chemical affinity, chemists tried to find a force similar to gravity in chemical reactions. They would not find affinity, but they would find other important mathematical models in the course of their search. We have much to thank Newton for: He legitimized the idea of force at a distance and he gave us many of the tools needed to build the theory of physical chemistry. And when he was done he went back to alchemy.

Why was Sir Isaac Newton interested in alchemy? For one reason: It tasked him. Mathematics and physics succumbed to his wonderful mind, so why not chemistry? He knew that Boyle had confirmed the mathematical relationship between the volume of a gas and the pressure exerted on it. Surely, Newton must have believed, if he tried hard enough he could apply mathematics to the mysteries of potions, too. So in a way, our consideration of Newton's efforts in the realm of alchemy points out why physical chemistry emerged as a distinct endeavor and why it continues to be an active area of research: Atoms and molecules are tiny and tricky creatures, difficult to tame. The physics of the planets is awesome, indeed, but so is the physics of atoms. Newton had found many keys that would be important to chemistry, but he would not be the one to unlock its secrets.

In the end, alchemy truly turned against him. It can never be known for certain if he suffered from mercury poisoning as a result of his experiments, but it is known that his laboratory caught fire in 1692 and his notes burned. In 1693 he suffered what would now be termed a nervous breakdown. In his apparent paranoia he accused several friends of various conspiracies against him, but the majority of his associates recognized the stress he was under and did not abandon him.

In 1696 Newton accepted the government position of Warden of the Mint and at the mint, Newton applied his knowledge of chemistry to assaying coinage metals. However, his role was supervisory and he spent his time at a writing table rather than a laboratory bench. He continued to publish, but became embroiled in an unfortunate debate over the origin of the calculus that lasted until Leibniz's death. However, he became president of the Royal Society in 1703 and held the post for the rest of his life. Queen Anne knighted him in 1705. Not a bad conclusion.

So in our consideration of Newton we have seen how the many concepts from the ancients proved fruitful, but alchemy was mostly a dead end. The alchemists stumbled on a couple of new reactions to add to the collection, but contributed very little in the way of real explanation, as Newton was finally forced to conclude. But the alchemists are not to be blamed. Knowing what we know now, we can safely say there was no way they could have made progress based on the scant information and tools they had at hand. But the tools and information were accumulating and coming from many diverse areas of human endeavor: Newton looked for codification of chemistry in alchemy, but Lavoisier found it in accounting.

Chapter 10

Lavoisier

THE CHEMICAL TSUNAMI

This theory is not, as I have heard said, the theory of the French Chemists, it is mine . . .

Antoine Laurent Lavoisier, circa 1790

In the mid-1700s the wave of the Scientific Revolution engulfed chemistry. As with all revolutions, it was the product of many minds, but we focus on primarily one—Antoine Lavoisier—as he would have wished. The difference between chemistry before and after Lavoisier is the difference between a story and a history. A story starts, "Once up a time . . ." A history starts, "In 1789 . . ." Lavoisier changed chemistry from a qualitative art to a quantitative science.

Before Lavoisier, European chemistry was a morass of confusing and conflicting myth and mysticism. The chemists were painfully aware of Newton's beautiful codification of physics and were groping for a unifying principle of their own. It all started when Newton's *Principia* was translated into French by the Marquise du Châtelet. Although Lavoisier came to be the hero, a few words need to be added about the marquise for two reasons: She deserves more than a footnote, and her life helps set the scene that was France when Lavoisier stepped onto the stage.

By the time of du Châtelet, our old friends the medieval merchants, who so happily supported the mathematicians, had become wealthy and influential to the point that they now challenged the

Church and the aristocrats for power. In France this new middle class was called the bourgeoisie and the bourgeoisie that bred Lavoisier bred du Châtelet, too. The parents of Gabrielle-Émilie le Tonnelier de Breteuil, Marquise du Châtelet (1706–1749) thought her too tall and plain to marry and not suited to convent life, so they allowed her to follow her natural inclinations and receive an education.[1] She married after all, a military man fifteen years her senior, who was often absent. The situation agreed with her as it allowed time to fraternize with the scientists and philosophers of the Enlightenment. It was through these associations that she met Voltaire, who became her lover and intellectual companion for some sixteen years. She was in the midst of translating the *Principia* when she found herself expecting her fourth child. She increased her pace in fear that she might not survive the birth, as many did not, and so managed to complete the translation and a commentary before giving birth to a son. She died six days later.[2] In happier times du Châtelet and Voltaire had built a rude chemistry laboratory in her chalet to investigate what was then the burning question of chemistry: the nature of fire. This question was to fascinate Lavoisier, too.

Lavoisier (1743–1794) was six years old when du Châtelet's translation of the *Principia* was published, and there can be little doubt that he eventually read it.[3] Coming from a wealthy family, Lavoisier was highly educated. As members of the French bourgeoisie, his family knew that their position was maintained by talent rather than title. Though Lavoisier would be a millionaire by today's standards, he lived his whole life as though the wolf were at the door.[4] As it turned out, it wouldn't be the wolves that got him.

Lavoisier's family had originally thought that he might be a lawyer, so he was trained in law and letters as well as accounting, but they also allowed him to indulge his talent in mathematics and science. With the deeply ingrained lust of the bourgeoisie for fame as well as fortune, Lavoisier aspired to the position of paid researcher at the French Academy of Sciences, an early prestigious think tank founded by the French monarchy, but this position promised an annual income of only some 2,000 livres (about $80,000 in 1996 U.S. dollars).[5] To support the lifestyle to which he had become accustomed, Lavoisier supplemented his income by investing, primarily in the French tax farm, an eventual fatal mistake.

The fatality of his mistake, however, was certainly not immediately apparent. A true descendent of the medieval merchants, the balance sheet was always the sacred and ultimate logic for Lavoisier, and for him the tax farm seemed a natural investment. The tax farm was an institution, not uncommon in European governments of the time, whose function was to collect taxes for the monarch. The profit was the difference between the amount owed the monarch and the amount collected. This situation lent itself to abuse at the very least, and in France the system was particularly ugly. Part of Lavoisier's duties as a tax farmer was to assess for tobacco and alcohol tax and to suppress smuggling of these commodities. The majority of French people at the time were quite poor, so smuggling was rampant. Tens of thousands of arrests were made annually and the majority of those arrested were children. The punishments were cruel, including whippings, hangings, or being left to starve in cages.[6] Lavoisier does not appear to have been a particularly harsh or unfeeling man, but it is apparent that he was first and foremost an accountant, and the books won out in any arguments he might have had with himself. In addition, his investment netted him around $3 million a year (in 1996 U.S. dollars). It also netted him two other factors that would be pivotal in his chemical career: a wife and a governmental appointment as the administrator of the gunpowder arsenal in Paris.

Marie Lavoisier's place in her husband's life can be concluded from numerous drawings of the laboratory done by her and the recorded recollections by friends of her participation in his work in the laboratory.[7] A daughter of a fellow tax farmer, she was thirteen when she married the twenty-eight-year-old Lavoisier. She came to the marriage with an education in languages. She learned chemistry from Lavoisier and their colleagues. Like du Châtelet, she had a talent for language, so she translated books and papers for Lavoisier, who was much better with numbers than foreign verbs. The marriage suited Lavoisier: He both realized a very generous dowry and gained an able assistant. The marriage also suited Marie Lavoisier. Again, like du Châtelet, she didn't mind that her older husband was often absent on inspection tours for the tax farm. She, too, was content to hold intellectual salons in his absence and eventually also took a lover: Pierre Samuel du Pont, the father of Eleuthère Irénée du Pont of E. I. du Pont de Nemours & Company, the American gunpowder manufacturer. The association was not coincidental. In the laboratories of the arsenal, Antoine Lavoisier had taught du Pont the gunpowder trade. If Lavoisier knew of the relationship, he did not reveal his knowledge. For Lavoisier, the balance sheet reigned supreme.

Sometime early on in his life it must have occurred to Lavoisier that the very same methods that were so useful to him in business were also his ticket to fame and fortune in the chemical arts. By 1774 Lavoisier developed a device, a mass balance, that could measure one one-hundredth the mass of a drop of water. He employed this balance, and a balance sheet, to revolutionize chemistry. One of his first assignments at the Academy of Sciences was to help settle a debate over the best way to measure the purity of water in order to find the best water source for Paris. The two choices were to look at the density of the water, which would change with the amount of dissolved impurities (just as a cup of gravy weighs more than a cup of pure water), or to evaporate the water to see how much scum was left in the pot. Lavoisier advocated the use of density as an indicator of purity because he thought some of the impurities could be carried off by the water as it evaporated. He was right. (That's why it is wise to boil drinking water of uncertain quality. Boiling kills bacteria and drives off volatile impurities.) Robert Boyle, as well as others, had observed that when water is subjected to repeated distillation, some solid always seemed to be left in the still. Aristotle had proposed that all matter is composed of four elements, fire, air, earth, and water, and alchemists and others had long held the belief that there should be a way of changing one material into another by manipulation of these four basic elements, for instance changing lead into gold in a process called transmutation. So others who had found residue in their stills took it to be evidence of a transmutation that had taken place while heating the pot. But Lavoisier was a believer in the balance sheet and needed proof. He boiled water for a hundred days in a sealed glass container. He found that, yes, there was a residue in the water after boiling, but when he weighed his water, residue, and container he found the same total mass as before the boiling. Where had the residue come from? When he weighed the container separately he found it had lost weight, the same weight as the residue. The credit and debit columns matched. The water had not transmuted into earth—prolonged boiling had resulted in dissolution of some of the glass container.

Having consigned transmutation to fable with his balance sheet, Lavoisier proceeded to establish the principle of conservation of mass. This principle can be illustrated by a simple thought experiment: If a dirty child is put in a bathtub of soapy water, a solid called scum is often observed to form and accumulate on the side of the tub. Clearly a chemical reaction has occurred because a new substance, scum, has been formed. But one would not expect the mass of the tub of soapy water plus child to have changed. The system—water, soap, and dirt—would change form, but mass would be conserved. With several experiments not much different than this one (but with chemicals, not children), Lavoisier decided that mass cannot change over the course of a chemical reaction.[8] If all the products could be contained and collected, then the total mass before and after the reaction must be equal. The elements may regroup into new forms, but none of the original elements are lost and no new ones are gained. No miracles. No magic. Rearrangement of materials, but conservation of mass. This one principle served to bring chemistry from the Dark Ages to the Enlightenment. Chemists could now hope to predict and calculate the output of chemical reactions. The medieval recipes of a pinch of this and a pinch of that could now become procedures for production.

Armed with his new principle, Lavoisier went on to tackle calcination reactions, which is the type of reaction that occurs when an empty tin or aluminum can is tossed into a campfire. The shiny, malleable metal becomes a gray, grainy, and fragile solid. Lavoisier established that when this reaction occurs in air, the mass of the solid formed is more than the mass of the metal before the reaction, and he wanted to know why. When he came up with his answer, it would not only sound the death knell for a quaint notion called *phlogiston*, it would also consign Aristotle's four elements to the dustbin.

It started simply as the difference between good air and bad air. Good air was the type of air that could support respiration and combustion. Bad air could not. The chemists could show the difference by putting a mouse or a candle under a bell jar and noting that when the candle went out or the mouse expired, there was still gas in the jar: water admitted into the jar would not fill it completely. The air that had been used up—the air that had kept the candle lit and the mouse alive—was the good air. The air left in the jar when the candle went out was the bad air. Then, in addition to the good air and the bad air, some investigators noticed that they could generate another type of air: *very* good air. Good air allowed a candle to burn, bad air put it out, but this new, *very* good air caused it to burn brighter. One of the English workers who noticed this new air, Joseph Priestley, was a self-educated scientist who had written extensively in the history of science. His reputation became such that he was hired as a librarian by an English noble, Lord Shelburne, and given the freedom to conduct experiments at Lord Shelburne's residence. On a trip accompanying Lord Shelburne to Paris, Priestley reported the new air to fellow scientists on the Continent. He called his new air *dephlogisticated air* because it was thought to be lacking *phlogiston*—a hypothetical material supposed to be given off during combustion. Dephlogisticated air, he proposed, promoted combustion by absorbing phlogiston from heated material. Lavoisier was having a hard time accepting phlogiston because the theory contradicted *his* principle: the conservation of mass. If heated materials lost phlogiston, why did calcined materials get heavier?

In the course of a conversation shared with Lord Shelburne and his continental hosts, Priestley mentioned his discovery to Antoine Lavoisier. How, asked Lavoisier, might one obtain a sample of this new gas? Priestley explained it could be obtained from heating a certain mercury salt. Where, asked Lavoisier, might one obtain a sample of this mercury salt? Priestley explained that it could be simply made by heating mercury metal in air.

This exchange of information is said to have taken place at a dinner meeting. It is amazing that the other diners were not blinded by the lightbulbs going off in Lavoisier's head: The gas given off by the mercury salt must be the same as the gas absorbed by mercury metal from the air. The mysterious phlogiston? Aristotle's element of fire? Aristotle's air? No, a pure part of air that is different from other parts: a new element. But Lavoisier did not shout out his revelation. He, unlike Priestley, played his cards a bit closer to his chest. He wanted this discovery for himself.

Lavoisier did an experiment with tin. He heated tin in a sealed container and found it turned into a white powder (as a tin can will do in a campfire) and the container and contents weighed the same before as after the reaction. But when he broke the vessel open after the reaction, air rushed in. He concluded that some component of the air had been fixed onto the tin in the reaction, just as it fixed onto mercury.

The pieces began to come together. Lavoisier experimented with other reactions that would not proceed in the absence of air, such as the combustion of diamonds. This curiously extravagant experiment had its origins in practicality. People who worked with diamonds— jewelers, diamond merchants—knew that when they heated diamonds to remove certain blemishes they had to wrap them in cement and enclose them in crucibles sealed as tightly as possible: If any air got into the diamonds they could char black. Lavoisier burned some diamonds in a container with air and found that the residue in the sealed container weighed the same as the contents before burning, but the collected diamond dust had gained weight. Lavoisier burned phosphorus and sulfur to form acid and found they gained in weight. He then pronounced his theory: Air, once thought a primary element, was actually composed of more elementary parts. One of these parts was the very good air of Priestley. But Lavoisier claimed its discovery for himself and named it "acid former," which in Greek is "oxygen." In short, he showed that Aristotle was wrong. There were more than four elements, or even five. In fact, Lavoisier soon declared there to be some thirty-three in all.

But why was this revolutionary? The four elements of Aristotle had been abandoned before: Meurdrac, in imitation of Paracelsus, had pointed to sulfur, mercury, and salt as being the primary elements. The revolutionary difference in the elements of Lavoisier was that his elements could stand alone. Aristotle, Meurdrac, Paracelsus, and others up until this time had been speaking of elements as the essential components of *all* matter: They thought that a little bit of each must be found in all material. Different materials were the result of different relative amounts of the basic elements. Lavoisier said, no, he could purify and isolate at least one pure element—then he turned his attention to others. He revisited and reinterpreted the results of other workers in terms of his new theory (though rarely acknowledging their contributions). He isolated a gas Priestley called "inflammable air," a gas that formed water when sparked with oxygen, and called it hydrogen, another element. With a lawyer's appreciation of language, he realized his new system of chemistry would have to have a systematic vocabulary to describe it. When he wrote a textbook, *An Elementary Treatise on Chemistry*, in which he described his thirty-three basic elements, he included an entire system of nomenclature.

The triumph was complete. Lavoisier became the premier authority on science and chemistry in France. He was named director of the Academy of Sciences in 1785. He was made a member of the commission on weights and measures in 1790. He became commissioner of the treasury in 1791.[9] Then, in 1794, Lavoisier's progress was interrupted.

In Lavoisier's time there were essentially three powers in France: the monarchy, the Church, and the bourgeoisie. The monarchy, however, which required taxes from and collected by the bourgeoisie, was in financial difficulty. The bourgeoisie, being merchants and book-keepers by definition, thought that they might be able to handle finances better and, as a member of the bourgeoisie, Lavoisier was initially in favor of the French Revolution of 1789. He and others believed it would be settled fairly bloodlessly by the institution of a parliamentary form of government.

But the change was not rapid or drastic enough to satisfy the populace. Conditions for the common peasant were brutish. There wasn't enough food to go around and punishments meted out by local nobles could be unspeakably cruel. A popular revolt arose that threatened to completely destroy the governmental infrastructure, and this drastic turn undermined foreign support for the revolution. There was widespread fear of foreign war, which did indeed result. In the ensuing panic the king was executed. Soon thereafter the queen was executed, and there was a general breakdown in economic, governmental, and judicial systems. Without regard to how one feels about Lavoisier's rather cavalier attitude toward the contributions of others and his seeming disregard for suffering, one may feel some sympathy for the man who made system his life and was now the victim of chaos.

Predictably—to everyone but Lavoisier—Lavoisier was arrested. No doubt the actual reason for his arrest was that he had been part of the hated tax farm, but the revolutionary court, going through the formality of trying him for a definite offense, charged him with falsifying the books for the tax farm and allowing tobacco to be adulterated. He was warned of the impending arrest and could have avoided capture, as many did, but his father-in-law refused to flee so Lavoisier likewise stayed and they turned themselves in together. In detention, while awaiting trial, he continued work on the books for the tax farm in the evident belief that he could explain to the tribunal each of their objections and show them that the charges were groundless. Unfortunately his wife felt the same way. It has been said that she was too arrogant when she made her plea for his life, that if she had been more supplicant or offered bribes she might have been able to free him. But if she were too arrogant, she was also too logical. She did not see why he had been arrested in the first place and said so.[10]

What was Lavoisier's defense to the charge of the tax farm's adulteration of tobacco with ash? He simply stated that he was aware of the addition of ash to tobacco and that in his experience he had found that customers seemed to like a small amount of ash because it added a certain pungent flavor. He had even developed a test for it. A major component of wood ash is various carbonates, such as sodium bicarbonate, or baking soda. Lavoisier knew, as all kitchen chemists do, baking soda reacts with vinegar to form a vigorously foaming mixture. Lavoisier took advantage of this reaction to test for the presence of ash in tobacco. He found that he could estimate the amount of ash by the amount of effervesce produced when acid was

Fig. 10. "The Arrest of Lavoisier" (1876), by L. Langenmantel. Edgar Fahs Smith Collection, University of Pennsylvania Library.

poured on a sample. But he had carefully regulated the amount because too much ash lost customers. The defense was perfectly logical—and perfectly unacceptable. The execution by guillotine of the twenty-eight tax farmers took all of thirty-five minutes.

Was this a great loss to science? Yes. Was this a travesty of justice? Perhaps. His friends were roundly railed for cowardice by Marie Lavoisier for not coming to his defense (though some of them tried),[11] and although fear was certainly a component of their reticence, they had other reasons, too, such as believing that the tax farmers deserved what they got. The farmers were a rather cool-blooded lot.

In a bit of justice, the analysis of Lavoisier's life comes down, once more, to a debate of qualities over quantities. Whatever his unmeasurable contributions or detractions to social welfare may have been, his contributions to chemistry are clear and definable. He applied the methods of physics to chemistry. He showed, incontrovertibly, that mass is conserved over the course of a chemical reaction. He demonstrated that air is not one substance, but several, and that one component of air, oxygen, is essential to combustion. He redefined the notion of element from Aristotle's famous four that are found in everything to an unlimited number of pure materials that can, and must, stand alone. If there is a sadness to his ultimate demise, then there is a sadness to the end of those around him, too.

Marie Lavoisier was herself imprisoned for over two months,[12] but in the end only had property confiscated. She tried to demand repayments of old debts from Pierre Samuel du Pont,[13] but du Pont fled to the new United States, as part of the wave of refugees fleeing the French Revolution. Enough of the Lavoisier fortune was eventually returned so that she was able to regain something of her former lifestyle, and indeed she eventually married another man of science, with whom she no doubt hoped to share an interest. But her new husband, Count Rumford, had a different image of their life together and the marriage ended stormily four years after it began. She continued, however, to entertain intellectuals in her home.[14]

We cannot conclude a chapter on Lavoisier without noting that this is only part of his story. In the title of this chapter we refer to him as a chemical tsunami, but we haven't fully explained the allusion. He achieved an enormous body of work, both in science and in economics, which we have described only in part. He worked on many problems in chemistry and derived numerous brilliant and unique solutions. He designed methods for the analysis of organic materials, such as the materials that make up plant and animal tissue. He contributed to the theory of organic chemistry. He designed a systematic method for naming inorganic compounds, that is, nonorganic materials, and investigated many inorganic reactions. He delved into the nature and the measurement of heat—an area fundamental to physical chemistry. In a final bit of work immediately before his arrest and execution, he managed to demonstrate that metabolism is combustion, too. But the contribution we celebrate here, and the reason for his unique place in the history of physical chemistry, is that Lavoisier finished the job of legitimizing chemistry. After Lavoisier, chemists, whose art had evolved from that of witches and alchemists, could now walk with heads held high—despite Lavoisier losing his. Boyle had confirmed that mathematics applied to gases. Lavoisier showed that numbers applied to chemical transformations, too. Newton codified physics. Lavoisier codified chemistry—but then opened the door to a new conundrum: If the air is composed of various gases, then why don't the dense gases settle to the ground and why don't tall people faint? This question would be pondered by an English schoolteacher by the name of Dalton.

Part III

The First Atomic Wars

INTRODUCTION

> *I am never content until I have constructed a mechanical model of what I am studying. If I succeed in making one, I understand . . .*
>
> Lord Kelvin, circa 1850

The beginning of the nineteenth century found European scientists delighting in what was termed *the mechanical universe*: the idea that straightforward equations, such as those that described Newton's gravity and Lavoisier's conservation of mass, would eventually be found for all natural phenomena. This sense of control over nature was only reinforced when nature tried to take a little revenge. When London ran out of logs, England turned to coal, and as result acquired more equations.

Coal is not nearly as clean a fuel as wood. Coal produces a black, sticky smoke, instead of white, billowy clouds, and early coal mining was brutal. But coal mining can be credited with inspiring the machinist, James Watt, to tune up the steam engine and make it a viable tool: the tool that brought about the Industrial Revolution.

In this new industrial age, a collective return to realism took hold. The wig, powdered, plastered, and perfumed, went out of style and natural hair appeared in portraits. People all over, including the aca-

133

demics, rolled up their sleeves and got to work. The Newton effect—
the search for system and codification that so inspired Lavoisier—
inspired others. In the mid-1700s, Carl von Linné devised a classification system for plants and animals (and coined the term *Homo
sapiens* to classify one peculiarly troublesome mammal), and in the
1800s Charles Darwin organized a departure from the Garden of
Eden with his *Origin of Species*. By the end of the 1800s an obscure
Russian chemist, Dmitri Mendeleev, recognized the intriguing
Pythagorean periodic properties of the new elements and organized
some sixty-six elements into a periodic table.

Diseases were classified and codified according to symptoms, and the
approach to cause and cure began to be experimental rather than philosophical. Edward Jenner introduced the smallpox vaccination; Louis Pasteur and Robert Koch, using the now quantitative methods of chemistry,
established the germ theory of disease. Even psychology was put on a
more mathematical basis when Gustav Fechner founded psychophysics.

Fig. 11. Nineteenth-century laboratory of J. Bell and Company, London. William Henry
Hunt (1790–1864) is thought to be the artist. J. D. Murray is the engraver. Edgar Fahs
Smith Collection, University of Pennsylvania Library.

Systemization led to specialization. At the beginning of the 1800s "Physik" was generally taught at universities by the philosophical faculty and "Chemie" by the medical faculty. Lavoisier imbued chemistry with new respectability, and the Industrial Revolution reinforced its importance. Soon chemistry was not just taught in conjunction with medicine but as a subject in itself. The first journals of chemistry began to appear in the late 1700s, and these journals became important catalysts for chemical progress.[1] Unfortunately, they also sometimes became pedestals for the high priests of science who refused to give voice to, and sometimes silenced, the expression of important new ideas. Chemists and physicists, conveniently ignoring their ancestry of freewheeling mathematicians, magicians, and machinists, began to exclude people from their club. We will still see some successful amateur and self-taught scientists such as James Joule and Michael Faraday, but they, too, paid some price for their lack of pedigree.

With this new respectability came profitability, too. It became possible to work for a university and do experiments in a laboratory financed through students' fees rather than one's own pocket. But the gentleman-scholar would still retain a strong tradition. Well into the twentieth century we will find university professors paying for equipment on their own or hiring their own assistants.

In mentioning "gentleman" scholars, it must be added that there were, of course, also "gentlewoman" scholars, as we shall see, but for much of the 1800s, the role of women in European science remained implicit rather than explicit. Such was the case with Marie Lavoisier, and such would be the case for Marie Pasteur.[2] They were not usually formally acknowledged as collaborators. In the late 1800s there were some researchers in some universities that welcomed and encouraged the contributions of female students,[3] but the majority of the universities excluded women. A woman who wished to express her talents either had to work without benefit of formal appointment (that is, for free) or align herself with a husband, brother, or male colleague. But in the 1800s, the universities of Europe were exciting, vibrant places to be and the manipulations necessary to be associated with one must have seemed well worth the effort.

The competition that often made it nearly impossible for a woman made it very difficult for a man, too.[4] In Germany, there were three formal steps to becoming qualified for a university position: the acceptance of a second major piece of scholarship after the acceptance of the doctoral dissertation (a process that normally took six to ten additional years of work), the successful organization of a scholarly convention, and the survival of a disputation with faculty and subsequent public presentation of one's work. But even after this arduous process, one was allowed only to sit on one's hands, carrying out independent research as well as possible, and wait, hoping that one's reputation would engender an offer at a university. When and if a faculty decided they needed you, you would be "called." Some waited as long as twenty years to be called, and some, of course, were never called. While waiting, it was normal to give private lectures or to tutor under the auspices of a university and to collect fees directly from the students for these services as *Privatdozent*. If such a Privatdozent was never called to a full position, he (rarely she) might, out of appreciation for his work, at least be offered an associate professorship, which carried the title *Extraordinarius* and from which he would realize a small but steady salary. However, he (or she) also might manage to offend someone of the established scholarly community and thereby effectively end a career. In England, both teachers and students at Oxford and Cambridge had to accept the Thirty-nine Articles of Religion of the Church of England and persons of dissenting religions who refused were excluded.[5]

Remarkably, or perhaps some might say predictably, the first new hero in our story is one who operated completely outside these rigid societal structures, as his detractors were wont to point out. John Dalton was not a professor, but a mere schoolteacher, who was a dissenter, at that. The members of our next cast, the nineteenth-century investigators of heat, were mainly university professors, but they were intellectual dissenters in that they strove to apply the principles of physics and chemistry to medicine as well as inanimate objects, thus crossing swords with some of the medical establishment. The kinetic theory of gas would fit into the picture of the mechanical universe, though the protagonists in this story did not fit so neatly into their social worlds. Those that initially promoted the kinetic theory of gas ran so far outside the herd that they encountered great difficulty in getting their ideas published.

Then, just as the paradigm of the mechanical universe seemed firmly in place, it was jolted by the discovery that some phenomena were best understood in terms of statistical, probabilistic models rather than mechanical levers and gears. When statistical treatments were proposed to deal with a curious quantity called entropy, they were met with resistance. This resistance evolved into grudging acceptance when similar treatments were used to find evidence for the invisible, elusive atom.

Though acknowledgment of the existence of atoms would pose interesting questions and open more doors, reliance on statistical methods for their verification meant relinquishing the mechanical universe and accepting that in the future science would progress as it had in the past, through an unending cycle of guess, test, refine.

Chapter 11

Dalton's
Diminutive Friends

. . . each particle occupies the center of a comparatively large sphere, and
supports its dignity by keeping all the rest . . . at a respectful distance.

John Dalton, circa 1800

In the wake of Lavoisier came a flood of questions, among them
one most ancient: What is air? According to Lavoisier, air com-
prises several elements, not just one, and he captured two of these
elements—hydrogen and oxygen—and put them on display. Armed
with Lavoisier's balance and his emphasis on mass, his fellow scientists soon found hydrogen weighs less, volume for volume, than
oxygen. In other words, hydrogen is less dense, just as cooking oil is
less dense than water. But if this were so, asked a quiet Quaker
schoolteacher, then why doesn't air separate into layers, as oil separates from water? If this were true, John Dalton would have noticed;
he kept a notebook on weather for fifty-seven years, the last entry
being on the evening before his death.[1] And not once had he noticed
stratified air.

The answer Dalton gave was that the elements would separate if
they came in sheets—so it must be that they did not. Just as Lavoisier
the accountant understood the value of the balance sheet, Dalton the
schoolteacher understood the value of a good instructional device.

138

Imagine, said the schoolteacher, that air is composed of many individual, indivisible, particles of gases. Then they would be randomly distributed in the mix, like many colored marbles in a bag. Dalton called his marbles "atoms."

John Dalton (1766–1844) wasn't the first to propose the existence of atoms. The Greek thinkers Leucippus and Democritus (circa 500–400 B.C.E.) thought in terms of atoms (and provided us with the word), though Aristotle, a hundred years later, opposed the idea. Robert Boyle (circa 1650) was an atomist, that is, one who used atoms in his model of matter, although he preferred the word "corpuscle."[2] So why do scholars credit Dalton with the invention of modern atomic theory? Because, like Lavoisier, he backed up his theory with numbers. Leucippus, Democritus, and Boyle believed that one type of atom was the basis of all materials. Dalton's epiphany was to decide that the atoms of different elements were different, and what made them different was mass. An atom of oxygen, said Dalton, weighs more than an atom of hydrogen. This notion, combined with a limited set of assumptions, allowed him to assign a relative mass to a large set of elements, which in turn suggested answers for some puzzling observations, including one with the curious-sounding name of *stoichiometry.*

The stoichiometry of a chemical reaction is the recipe for the reaction. Just as a recipe for a cake may call for two cups of flour for every one cup of sugar, a recipe for sulfuric acid calls for 80 grams of sulfur trioxide for every 18 grams of water. The word *stoichiometry* (from the Greek word for element, *stoiceion*, and an ending meaning measurement, *-metry*) was invented by Jeremias Richter, and came into use after he and others began noticing that some materials seemed to always react in fixed, whole-number ratios.[3]

In 1808, after ruminating on Lavoisier's conservation of mass and Richter's ratios, John Dalton decided that all the elements must react in fixed whole-number ratios, and these ratios were the result of elements existing as atoms of characteristic atomic weight. Dalton published a 916-page treatise on the stoichiometric approach to chemistry, *A New System of Chemical Philosophy*, allotting five pages to his proposed atomic theory.[4]

John Dalton must have known that his theory would cause controversy, but he was accustomed to that. As may be recalled from the trials of our medieval magicians and mathematicians, the Protestant Reformation in Europe took several quite radical turns and resulted in violent persecutions and devastating wars. In England, the Reformation, despite periodic violence, was more subdued than on continental Europe. However, those not part of the English solution to the Reformation, the Anglican Church, found themselves on the outside looking in. Because of the link between the Church of England and many of the universities, many members of dissenting religions had to do without education or form schools of their own.[5] John Dalton, son of a Quaker weaver, was such a dissenter who was privately tutored and home educated. By the age of twelve he found himself giving lessons in the local Quaker meeting house. Teaching was obviously his forte, so when his cousin retired, he and his brother took over their cousin's school, where they taught languages and mathematics. At the age of twenty-seven, Dalton found himself professor of mathematics and natural philosophy at New College, a college for those of dissenting religions, in Manchester. The college grew and eventually relocated to a larger population center, but Dalton opted to stay where he was. His brother and friends were in Manchester, and Dalton, unmarried and free of urgent need of income, supported himself through private tutoring, in the fine tradition of many of our previous heroes.

John Dalton's exclusion from mainstream education may have been a blessing rather than an impediment. In the early 1800s, chemistry was not yet taught at Cambridge or Oxford, because it was not considered suitable for a gentleman's curriculum.[6] But Dalton, using Lavoisier's *Elements of Chemistry* as a textbook, taught, experimented, and speculated on the nature of chemistry, perfectly free to form his own notions and not feel the need to conform.

Perhaps it was this freedom that led him to reexamine Richter's data and formulate what is now known as the *law of multiple proportions*, which can be visualized as follows. Chemicals are like people: They react in simple, whole-number ratios. They can be a family of 5, a couple of 2, or even a mélange of 3, but a group of 2.5 would mean sawing someone in half, which is not acceptable. Why should chemicals display this peculiar proclivity? Because, said Dalton, they are composed of individual units called atoms, and the atoms of each element are indivisible and unique.

As such, this law was a pleasing speculation, but not much more, until Dalton took it one step further. He declared that if someone were to make a limited number of simplifying assumptions, then a table of relative weights could be deduced. Using Lavoisier's conservation of mass, we can see that if the mass of one atom were known, then the mass of the other atom could be calculated by subtracting from the whole. The masses of compounds could be found by adding together the masses of the atoms that formed them. Elements were beginning to act more like numbers all the time. Lavoisier would have smiled.

The first simplifying assumption that Dalton made was to suppose that if a single compound were to form between two elements, then it contained one atom of each element. This would be analogous to assuming that all marriages are monogamous. Unfortunately, this led to some erroneous results (for instance, the well-known formula for water, H_2O, was taken by Dalton to be HO because it was the simplest compound between oxygen and water). But Dalton's assumption was a start. Dalton also came up with a method for depicting atoms as circles with different symbols within them (straight line, cross, star, and so on) for the different elements, and in so doing joined Lavoisier in doing for chemistry what the Renaissance mathematicians had done for algebra: taking chemistry from rhetorical form into the symbolic stage, which made it easier to think about and systematize.

During the 1800s, several other experimental observations were made that supported Dalton's conclusions about the atomic nature of matter. Perhaps the first evidence, or maybe even inspiration, for his theory, came from some results of an industrialist/experimentalist friend, William Henry (1774–1836), who made his observations on the solubility of gases.[7]

Henry, who ran a chemical plant for the manufacture of milk of magnesia, was also a self-styled chemist, dedicated to careful experimentation. His hands had been crushed by a beam when he was a child so he suffered from chronic pain, but he turned the enforced inactivity into an opportunity to study and pursue an interest in medicine. He worked with Dalton to establish that methane and ethane were two separate compounds composed of a fixed ratio of 1:4 for carbon to hydrogen.

Henry also found that the solubility of a particular gas in a solution depends on the pressure of *that particular gas*, not the total pressure, a truism now known as Henry's law. In other words, if a solution has an atmosphere of nitrogen above it, some of the nitrogen will dissolve in the solution. This is why fish can breathe: Some of the oxygen in the atmosphere dissolves in the water in which they swim. However, only an added pressure of nitrogen gas above a solution can force more nitrogen into the solution; that is, mixing helium to a mixture of nitrogen gas above a solution may increase the total pressure above the solution, but will not increase the amount of nitrogen in solution. Thus, Henry's law is important to deep-sea divers who like to dive, but who do not want to get the bends.

The pressure of the air in the tanks carried by deep-sea divers has to equal the pressure of the outside water or the lungs will be crushed. But increasing the pressure of the tanks by adding more regular breathing air is not good. The air we breathe is mostly nitrogen with some oxygen mixed in, so if we increase the pressure of breathing air in the tanks, we are increasing the pressure of nitrogen. Increasing the pressure of nitrogen makes it more soluble in fatty tissue, which can lead to nitrogen narcosis or bubbles forming on ascent, the painful condition known as the bends. The answer to the problem is to increase the pressure in the tanks with helium: Helium will not force nitrogen into the fatty tissues; only nitrogen can force nitrogen into solution. Dalton saw this behavior of gases as added evidence for the individuality of atoms.

His ability for precision experimentation, which had allowed Henry to observe the solubility of gases, became limited as he grew older. The injuries to his hands began to interfere with his dexterity and finally caused such acute pain that it deprived him of sleep. Henry committed suicide at the age of sixty-two. But Dalton continued to find support, including from the team of Pierre Dulong and

Alexis Petit, who published their law of specific heats in 1819. Specific heat is the amount of heat required to raise the temperature of a given amount of material one degree on a temperature scale. Another name used for the same idea with different units is "heat capacity," a choice which better implies the meaning: the capacity of a material to hold heat. A material has a high heat capacity if a large amount of heat is required to raise the temperature; it holds more heat before the temperature goes up. The specific heats for different materials vary; for instance, the amount of heat required to raise the temperature of copper is less than the amount of heat required to raise the temperature of stainless steel. For this reason, the bottom of the best stainless steel cookware is copper coated: Heat is transferred more evenly to the sauce in the pot via the copper. The team of Dulong and Petit was part of the new wave of French chemists coming out of postrevolutionary France. The breakthroughs of Lavoisier had breathed new fire into French chemistry, and the enthusiasm and productivity of the French chemists was impressive. However, unlike Lavoisier, Pierre Dulong (1785–1838) struggled against poverty and illness.[8] Dulong was orphaned when he was four, but his parents left him enough funds to go to school. He practiced medicine for a while in the poorer part of Paris, but lost much of his inheritance overall because many of his patients could not pay.[9] He lost a finger and an eye in a laboratory explosion in the course of one research project but found his reward with another, the one for which his name was immortalized, along with that of his friend, the mathematician and physicist Alexis Petit (1791–1820).[10] Dulong and Petit found that the product of the atomic mass (which is what Dalton's relative mass is now called) times the specific heat was a constant for a large number of metals. Although we now know that this is only approximately true, at the time it gave good evidence that there was something to these atoms of Dalton's and his scale of relative weights.

But perhaps the most compelling arguments for the atomic theory came from organic chemistry—the lively field of chemistry that amuses itself with the oft ill-smelling and greasy carbon-based chemicals that compose the stuff of animals and plants. This field was undergoing a revival of its own during the 1800s because of the plenitude of new materials from coal tar and because the multitalented Lavoisier had contrived a way to measure the carbon, oxygen, and nitrogen content of organic compounds. His original methods were crude, giving only approximate values, but the chemists of the 1800s improved the methods until they were able to achieve quite accurate analyses. In the 1820s, through these methods, a German chemist, Friedrich Wöhler, analyzed silver cyanate and found it to be of the same composition as silver fulminate, analyzed by his friend, Justus Leibig. They found that the same number and same type of atoms are in each compound; but silver fulminate explodes and silver cyanate does not. The explanation was one more stroke in the tally for Dalton's atoms: These compounds functioned differently because of the way their atoms were arranged, just as the functionality of a car that has its wheels attached to the axle differs profoundly from one with its wheels attached to the roof. Limonene, a compound that lends odor to citric fruits, smells like oranges in one configuration, while its mirror image smells like lemons.[11]

But not everyone was happy with the atomic model. Physicists and chemists felt very uncomfortable postulating the existence of something they could not hold up to the light and examine. Many saw the atomic theory as a convenient device, just as the editor of Copernicus's book rationalized Copernicus's theory as a computational convenience for the determination of the date of Easter. And so, like Copernicus, Dalton had his enthusiasts, but had his detractors, too. By Dalton's time, they may have been beyond burning freethinkers at the stake, but they did roast him. Many scientists preferred the word "equivalents" to atoms in an effort to avoid the issue, but other, more outspoken chemists could be virulent. The strongest objection was perhaps most succinctly summarized by Ernst Mach when he said, "Have you ever seen one?"[12]

Although Dalton was awarded high honors in England (over the objections of the high clergy of the Church of England),[13] as Dalton aged, he found attacks on his theory more and more intolerable. Assuming a defensive posture, he began to resist even justifiable refinements to his model. Jöns Jakob Berzelius (1779–1848), a physician and chemist, had the audacity to find a mathematical error in

Drawn & Etch'd by J. Stephenson

John Dalton,

D.C.L. LL.D. F.R.S. L. & E. &c &c

President of the

Lit.y & Phil.l Soc.y Manch.r

Fig. 12. John Dalton. Edgar Fahs Smith Collection, University of Pennsylvania Library.

Dalton's *New System* and to propose that the atoms were held together by an electrical attraction.[14] Berzelius also suggested representing the elements with letters (in what later came to be the modern manner, e.g., H_2O) rather than circles and symbols. Dalton resented such re-interpretation and new representations of *his* atoms.

There were other points on which legitimate objections to Dalton's atomic theory could be based. For instance, there was the well known fact that gases fill balloons. The atoms that Dalton envisioned were static, much like marbles in a bag, and it is an equally well known fact that marbles will not spontaneously blow up a balloon. Dalton offered an explanation: He said that each atom was surrounded by an atmosphere of heat and when a material was heated, this blanket of heat increased, which held the atoms apart and even allowed them to fill up balloons in the gas phase. As it stood, this explanation was a perfectly acceptable visual aid that would certainly help a schoolteacher instruct students in the properties of different phases of matter, but it was not very useful as a scientific theory. Dalton's description of the atmosphere of heat was qualitative, not quantitative. It did not have a number associated with it. His theory of heat was a throwback to the "green cheese" approach to modeling: Saying that the moon is made of green cheese does explain its irregular appearance, but more concrete data is needed to discern its true composition. Yet clearly heat would be involved in the explanation because heat is what caused the transition from one phase to the other. To move on in their understanding, the scientists of the 1800s would have to pay their respects to Dalton, then regroup, reconsider, and re-address another ancient question: What is heat?

Chapter 12

Thermodynamics

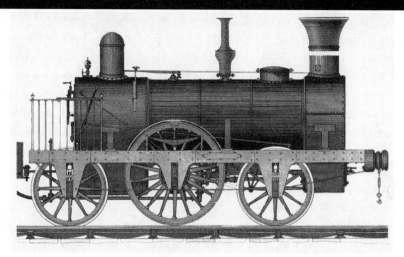

A clay pot sitting in the sun will always be a clay pot. It has to go through the white heat of the furnace to become porcelain.
Mildred Witte Struven, circa 1950

What is heat? We add heat to food to cook it, but we don't weigh it out or measure it with a spoon. We can create heat by rubbing our hands together, yet we can't seem to store it. Even in a thermos bottle, food eventually cools. Temperature is not the same as heat. Heat added to a metal spoon makes the spoon too hot to handle; the same amount of heat added to a wooden spoon allows it to be held comfortably. A hand in a freezer results in a cold hand because heat seems to flow downhill from where it is to where it isn't. But touching a chair results in no noticeable loss of heat, so the chair must contain heat. If the chair contained no heat, the first time you sat down would be your last. But as difficult as the concept of heat might be, in the 1800s there was an imperative to understand it. The heat of the steam engine, based on coal, was quickly changing the nature of societies worldwide.

The name of James Watt (1736–1819) is often associated with the steam engine, but James Watt did not invent the steam engine.[1] In fact, the steam engine is not even a nineteenth-century invention. Steam was used in the first century in Alexandria, Egypt, to power various mechanisms such as one used to open temple doors.[2] Thomas Newcomen (1663–1729), a blacksmith, reinvented the steam engine, but his steam engine was so inefficient that its only use was to pump water out of coal mines where there was plenty of poor quality coal laying around that could be used to power it and that would otherwise go to waste.[3] For this reason, running out of logs is identified as the catalyst for the Industrial Revolution rather than the reinvention of the steam engine: Without the need to remove water from coal mines, the steam engine would have remained a curiosity rather than an instrument of societal change.

A professor of natural philosophy at Glasgow in coal-rich Scotland, John Anderson, tried to demonstrate a small-scale model of Newcomen's engine in his lecture but couldn't get it to work. Anderson was a bit at odds with the university administration because he allowed artisans and other nonuniversity personnel into his classes (a practice branded "anti-toga"),[4] and to help him with the steam engine he enlisted the aid of another maverick, the university instrument maker, James Watt. Watt found himself at the university because the local guild would not let him practice in town without the requisite apprenticeship. In the course of Watt's efforts he not only got the model working but made it run better than its full-scale counterpart.

Understandably, Watt found himself fascinated with his better mousetrap and went on to study its workings. Watt found a way to assign a number to the amount of work done by a particular engine: He measured the pressure in the cylinder as it changed over the course of a power stroke. The product of the pressure times the volume expanded was a measure of the work done (as it takes more work to blow up a big balloon than a small one). In this way, Watt could compare the amount of work a particular engine extracted from a given amount of coal. In other words, he showed coal produced heat and coal produced work. Once it became apparent that work and heat were beasts of the same stripe, some pieces began to fall into place. This new animal, which encompasses all expressions of the ability to move or cause change, is called *energy*.

So now we see that some of our difficulties in understanding heat arise from imprecise language: We would be better off defining heat as the process of transferring energy from one body to the next and defining energy based on Dalton's atomic model of matter, but with an important modification. Dalton assumed that atoms were static, that is, atoms didn't move. Today we envision atoms in constant flight. It is the sum of this ceaseless motion of atoms that we call the energy content of a system—a concept that we can now use to tackle the difference between heating a system and raising the temperature.

Energy is transferred from atom to atom or molecule to molecule by collision: Atoms and molecules bang into one another as they fly around and, in the process, transfer energy back and forth. The game of billiards (or pool or snooker) is based on the idea of energy transfer: The cue stick transfers its energy to the cue ball, which strikes and transfers its energy to the object ball. The motion of the billiard ball on the table and the motion of atoms or molecules flying across the room is called translation motion. A rise in temperature occurs when the molecules of a substance transfer their energy to a thermometer through translational motion; in other words, by hitting the thermometer. The atoms of the mercury in the thermometer start moving around more and taking up more space. This expansion of the mercury in the thermometer is what registers as a change in temperature.

Now imagine that the corner pocket of a billiard table is a thermometer and the temperature goes up when an object goes into the pocket. Which object would more efficiently increase the temperature, billiard balls or jellyfish?

The obvious choice is billiard balls. Hard strikes on a billiard ball cause the ball to move forward vigorously. Even if the ball bounces randomly around the table, odds are it will eventually hit the corner pocket. Hard strikes on a jellyfish just cause the jellyfish to quiver and roll and move forward only sluggishly. Most of the energy is absorbed in the internal motion of the jellyfish, which doesn't get it any closer to the corner pocket.

An atom may behave like a billiard ball, but a molecule is more like a jellyfish. When energy is added to a molecule, it will quiver (vibrate) and roll (rotate) as well as translate. When energy is added to a molecular material the temperature does not go up as much as when it is added to an atomic material. This is why metal kitchen utensils have wooden handles. The metal transfers energy efficiently to the wood, but the wood is made up of molecules and pockets of molecular air that absorb energy in rotation and vibration instead of transferring it all to your hand. And that is why heat isn't the same thing as temperature: Adding heat to molecules doesn't raise the temperature as much as adding heat to atoms.

But none of this knowledge was known or even accessible to the early nineteenth-century scientists and therefore their accomplishments are that much more impressive. The number of great minds that engaged in the effort is just as impressive, but space forbids us naming them all. However, there are two fundamental principles in the mathematical model of heat, called the first and second law of thermodynamics, respectively, and we can use these as the focus of this chapter. The first law is that the energy of the universe is constant, and the second law is that the entropy of the universe is increasing. We will look at the conservation of energy first, and we will define and tackle entropy second.

Lavoisier thought the process of heating a material consisted of adding a substance, called *caloric*, to the material. This notion of heat as a substance would by challenged, interestingly enough, by the man who married Lavoisier's widow: an American-born soldier of fortune, Benjamin Thompson, who was honored by the Holy Roman Empire with the title Count Rumford. He and Marie Lavoisier were fellow beneficiaries of the gunpowder trade, but the explosive mixture only lasted a few tumultuous years, as did the marriage.[5]

Thompson's place in the history of heat is due to observations he made while boring out metal barrels for use as cannons. He found that as long as he kept boring the cannon, the cannon kept heating up, thus putting the idea of heat as a substance, caloric, into question. He asserted that heat was the transfer of motion: The motion of the cannon bore caused motion in the cannon, which heated the cannon.

Though Thompson's argument for heat as the transfer of motion was compelling, it didn't immediately carry the day because it didn't answer all the questions. For instance, although some transfer of energy might be explained by contact, the sun's light will warm an object through the vacuum of space. This difficulty was dealt with by distinguishing between conduction and radiative heating: Radiative heating is fast and conductive heating is slow. Because steam engines run well both on sunny and rainy days, radiative heating was moved to the back burner during the Industrial Revolution.[6] In an attempt, then, to show that conductive heating was caused by the transfer of motion, James Prescott Joule (1818–1889) demonstrated that a tub of water could be heated by a paddle wheel.[7] He measured the amount of work done by the paddle wheel by powering it with a weight dropped from a measured height. In this way, he arrived at a numerical equivalence between work and heat, which took the concept from the realm of speculation and put it on the bookshelf of science. In his honor, a unit of energy, the *joule*, is named after him. A similar unit of energy is the calorie. When we speak of a food containing some number of calories, we could as easily say it contains some number of joules.

James Prescott Joule was born into a well-off family of brewers in Manchester. As a young man he was tutored by John Dalton and he shared with his mentor a love for basic science and an obligation to find altruistic outlets for his talents. Having been to Edinburg, he was aware of the lethal byproducts of coal, so he took as his aim the improvement of electric motors, hoping to replace steam engines with electric ones. He used a laboratory he set up in his father's home to show that current passed through a liquid causes the temperature of the liquid to rise and then in 1850 he used his paddle wheel to show that mechanical movement causes temperature to rise, too.

At first Joule received a cool reception from the scientific establishment. He did not hold a position in the scientific community and he was not even a university graduate. However, his work gained the recognition of one William Thomson, better known as Lord Kelvin (1824–1907). Kelvin was a member of the inner circle, and after discussions with Joule, and an accidental meeting with him in which he found Joule carefully measuring the temperature at the top and bottom of a waterfall, he decided this young man had something to offer, and backed Joule with his own respected reputation.[8]

Kelvin's place in the history of heat is a more difficult one to decipher. He devised an absolute temperature scale that circumvented the problem of the zero and negative temperatures present in Celsius's scale, but this was really a revival of an earlier idea put forth by Guillaume Amontons (1663–1705).[9] He produced a huge body of work but is noted for his computational errors and erroneous conclusions as much as for any one unique insight. He is sometimes credited with coining the word "thermodynamics" (*thermo* for heat and *dynamics* for its movement) that became the name of heat science, but William Rankine probably used the word before him. Then why is Kelvin always included in the history of thermodynamics? Because he had the good sense to recognize and support sound scientific effort whatever the source. Without Kelvin, Joule would have faced more adversity.

But Kelvin could not shield Joule from all the cruelties that would befall him. Joule's wife died young, leaving him with two small children. The monetary burden of raising these children coupled with the expense of personally financing his research used up his inheritance. He spent his last years poor and sick. There were other misfortunes in Joule's life, too. He believed that not only were electrical energy and mechanical energy equivalent, but that all forms of energy could be interconverted, which meant that energy was not created or destroyed in a process, but conserved. However, his claim to be the first to state the principle of conservation of energy was disputed by others. The irony is that the basic idea had been around since Descartes and Newton. Descartes postulated a conservation of motion so that the creator wouldn't constantly be called upon to wind things up. Newton's third law, that for every action there is an equal and opposite reaction, is a statement of his intuitions regarding the conservation of energy. Although Joule and others had the idea, they didn't have the equations to support it. The mathematical formulation of the conservation of energy would come from a German doctor of medicine, Hermann Ludwig Ferdinand von Helmholtz (1821–1894).[10]

The mother of Helmholtz was a reserved daughter of a German artillery officer. His father was a Romantic philosopher who shared music, painting, and Kantian philosophy with his son. His parents did not have the money to send their son to a university, so they encouraged him to seek support from the state to study medicine. Helmholtz's approach to medicine was analytical, which now is the accepted, indeed expected, approach to medicine. But in the mid-1800s, it was not. In a movement that came to be known as the "1847 school of physiology," Helmholtz and others rejected the belief that there was something special or impenetrable about biological processes and insisted on grounding physiology firmly in the tangible principles of chemistry and physics. He even suffered a break with his father over the rigidity of his approach. Helmholtz's life was marred by other tragedy: His wife suffered ill heath and died young, leaving him with young children. He endured personal attacks from colleagues offended by his placement of chemistry and physics alongside anatomy in medicine. But it was his command of physics that allowed him to sort out the various expressions of energy and show their equivalence through mathematical form. His mathematics clearly stated and demonstrated the first law of thermodynamics: The energy of the universe is conserved.

The meaning of the first law may again be described by billiards. When a cue ball strikes an object ball, the object ball moves off, but the cue ball is also affected: It stops or at least slows down. Some (or all) of the energy of the cue ball has transferred to the object ball, but the net amount of energy contained in the system has not changed. (This concept will turn out to be very important in our later consideration of chemical thermodynamics.) In his formulation of the first law, Helmholtz read and drew from Joule, Kelvin, and others, including a quiet French military engineer, Sadi Carnot (1796–1832). Carnot deserves mention for his groundbreaking analysis of heat, but his unfortunate early death from cholera stopped him just short of formulating the second law of thermodynamics, that entropy is increasing.[11] This principle would be enunciated by another German, Rudolf Julian Emmanuel Clausius.

The studies by Clausius (1822–1888) of the work of Joule, Kelvin, and Carnot culminated in the publication in 1850 of a paper that makes a definitive statement of the law of conservation of energy, but does not stop there.[12] Clausius had ruminated on an observation made by Carnot: All real engines are less than 100 percent efficient. It is never true that all of the energy added to a system is completely converted to work. Some of the heating is wasted. Clausius worked another fifteen years on refining this concept and understanding its implications before he gave it a name: entropy.

Fig. 13. Hermann von Helmholtz. Edgar Fahs Smith Collection, University of Pennsylvania Library.

We now think of entropy as the tendency of all systems to go to a state of maximum disorder. A deck of cards becomes more mixed when shuffled; it does not spontaneously re-order by suit and rank. A sack of red and green marbles exists as a uniform mixture of marbles; it will not suddenly separate into a layer of red and a layer of green. Systems mix spontaneously: A few drops of blood will color a basin of water red. These observations are more than just common experience, they are driven by the same force that drives chemical and physical change. Heat spontaneously spreads from hot bodies to cold bodies because that is the direction that causes the most disorder. A box of warm, jumping frogs is more chaotic than a box of cold frogs huddled in the corner. These thoughts were summarized in Clausius's statement of the second law of thermodynamics: The entropy of the universe is increasing. The energy put into a system cannot all be converted to work because that would be too orderly. Nature insists on a bit of disorder, too. Entropy is the tax we must pay for borrowing nature's power.[13]

The second law, as well as the first, is of major importance to physical chemistry. It is the two tugs of entropy and energy that determine which reactions will occur spontaneously and the extent to which they will occur, as we will see when we examine these concepts more closely. Entropy has its effect on Clausius's life, too. Of Prussian birth, he carried out thermodynamic research at the University of Zürich, but was a German patriot all his life. In the German–French War of 1870, at fifty years of age, he organized and drove in an ambulance corps. The venture resulted in a leg wound that plagued him the rest of his life. In 1875 his wife died in childbirth, leaving him the job of raising their children, a task which he undertook personally, to the detriment of his professional duties. Even so, two of his six children died before maturity. After his children were raised, however, Clausius still did not devote much energy to formulating a mathematical model for the molecular basis for entropy, which he knew must exist. In the end, he deferred this advancement to an Austrian physicist whose intellectual roots lay with the carousing medieval mathematicians. But perhaps Clausius did this with no regret. Perhaps he anticipated the battles Ludwig Boltzmann would have to face. Perhaps Clausius preferred to end his life in peace.

Chapter 13

Gnats in Sunbeams

. . . a vast multitude of small particles of matter, traversing backwards and forwards in every direction . . . a swarm of gnats in a sunbeam.
John James Waterston, circa 1840

The molecular theory of entropy did not spring full grown from the brow of Boltzmann; therefore, it is fitting, as well as instructive, to pay homage to the theoretical mavericks who paved his way. These theorists had to travel outside the herd because the herd, in the main, trailed after Newton, and Newton assumed that atoms stood still. Newton had won his following by deriving Boyle's law—that the pressure of a gas goes up when the volume goes down and visa versa—by assuming the particles of matter were static but they exerted a repulsive force on their neighbors. Because of Newton's mathematical treatment and his stature, even Dalton, with his epiphany of atomic weight, still considered his atoms to be standing still. There were those, however, who were not comfortable with Newton's almost mystical action at a distance, which led them to explore other possibilities. A reasonable alternative was a model in which atoms were in constant motion and creating pressure by bumping each other out of the way. But the predictions of a viable theory must match experiment, and therein lies the story.

One of the earliest versions of this model, which was to become known as the kinetic theory of gases, was proposed by Daniel Bernoulli (1700–1782), a son of the famous Bernoulli family that

157

produced a pack of mathematicians, physicists, and scientists of various stripes in the 1600s and 1700s.[1] In the competitive tradition that seems so unique to early European mathematicians, Daniel Bernoulli even found himself pitted against his father, Johann Bernoulli, in mathematical contests. But when it comes to priority in the kinetic theory of gases, Daniel won the claim.[2]

By kinetic, we mean motion, as in kinetic art, the art that produces mobiles. The kinetic theory of gases takes the particles of gas—atoms or molecules—and predicts physical properties by assuming they are in motion. The theory is attractive because it, like Newton's, predicts Boyle's law, and in the hands of Bernoulli, made predictions beyond Boyle's law as well.

To understand the kinetic theory, we must put ourselves in the mindset of the theorists. They considered a collection of particles of gas to be like a swarm of gnats in a container, constantly moving in all possible, random directions. They recognized that the collective force of these gnats, continually striking an area of their container, would result in a pressure. If the container were made smaller, then the gnats would strike the area more often, and the pressure would go up: Boyle's law. If the swarm of gnats were heated, they would fly around faster and there would be more collisions with the walls of the container. If these walls were movable (e.g., a balloon) the volume would increase, as was observed by the famous French hot-air balloonist of the late 1700s, Joseph Gay-Lussac: increasing the temperature increases the volume of a gas.

But interesting as these results were, they still agreed only qualitatively with the findings of Boyle and Gay-Lussac: They didn't lead to a prediction that could be measured in any tangible way. To have this kind of comparison, the atomic theory, with its many different types of atoms, was necessary. For the Bernoulli father and son, all air atoms were the same. John Herapath, on the other hand, had the advantage of believing in different types of atoms in each different type of gas . . . not that it did him any good. The lesson to be learned from John Herapath is that if you want to be a famous scientist, take care that you are not born in the wrong place or at the wrong time: a caution John Herapath did not heed.[3]

English-born John Herapath (1790–1868) had some introduction to the principles of chemistry by virtue of being born to a family whose business it was to process malt. But Herapath had little formal education. He taught himself enough French to read the works of the French mathematicians and gained a local reputation as somewhat of a mathematics prodigy, enough so that he was able to establish himself as a teacher of mathematics. Then, apparently without knowledge of Bernoulli's previous work, he derived a kinetic theory of gases. He wrote it up (along with a treatment of gravity and heat) and submitted it for publication to the Royal Society. A certain Sir Humphry Davy (quite prominent in the annals of chemistry for his discoveries of elements and chemical reactivities) was then president of the Royal Society, and though Davy was a remarkably talented chemist, he wasn't much of a mathematician, and rather than struggle with the paper, he deemed it inappropriate for publication.

It was true that there were some difficulties with Herapath's treatment and some out-and-out errors, though these were not Davy's grounds for rejection. Herapath, with some justification, saw the rejection as an attack on his acumen. Herapath protested, and the ensuing exchange accelerated to the point of Herapath publishing a two-column tirade against Davy in the *London Times* and Davy eventually resigning the presidency of the Royal Society, ostensibly ill but probably to avoid a challenge to a mathematical duel (shades of medieval mathematicians) that Herapath proposed.

Herapath did eventually see the paper published in the *Annals of Philosophy* (after cleaning up some of the errors) but he would never gain the recognition he believed he deserved. In an interesting turn of events he became a contributor, then editor, of a publication called the *Railway Magazine,* and using as his justification the fact that trains travel through air and ice freezes on rails, he deemed a theoretical treatment of air pressure and temperature as entirely appropriate for a railroad magazine and used it as a venue for publication.

If Herapath made errors in calculations, he could be excused for not catching them because he didn't have a lot of experimental data with which to test his theory. This situation gradually changed through the 1800s. Scottish-born Thomas Graham (1805–1869)

Fig. 14. Daniel Bernoulli. Edgar Fahs Smith Collection, University of Pennsylvania Library.

added some information to the data bank by his study of gas effusion, the process by which gas escapes through a pinhole orifice.[4] He demonstrated in 1833 that the rate of effusion for gases depends inversely on the square root of the mass. In other words, helium, a very light gas, effuses more rapidly than nitrogen, a heavier gas that is the main component of air: A helium balloon will deflate faster than a balloon blown up with air because helium will effuse faster through the microscopic pores in the balloon.

Graham was so impassioned of the study of chemistry that he pursued it even though his father frowned on his career and withdrew financial support from his son. He supported himself by writing and tutoring (which have been the two stop-gap careers of many of our heroes) and Graham's chemistry was decidedly physical. He studied gases, explosions, and colloids—the interesting bigger-than-molecules but smaller-than-solids particles that would later be germane to twentieth-century physical chemistry. In a clever, if decidedly delicate experiment, Graham was able to show that, at a given temperature, the rate at which gases seep out of a pinhole (the rate of effusion) goes up as the square root of the mass goes down. His finding that the rate of effusion should be inversely proportional to mass is intuitive. Even if one does not accept the atomic theory, it is commonly understood that heavier objects move more sluggishly than lighter ones (the same kick imparts more movement to a tennis ball than a bowling ball). And the faster the gas particles move around, the greater their chances of finding the pinhole and escaping. But that the rate of effusion should be inversely proportional to the *square root* of mass has no explanation until the kinetic theory is evoked.

Graham's observations gave the kinetic theorists something that their theories had to account for, but they still were lacking some real numbers to compare with their calculations. Our next protagonist, John Waterston, recognized, as did others after him, that when absolute numbers aren't available, ratios can work as well. This is the way Dalton contended with weighing his minuscule atoms: He knew he couldn't assign absolute weights (it is hard to weigh one gnat, especially one you can't see), so he assigned relative weights by finding the ratio of weights of atoms. Waterston did the same, but with ratios of heat capacities, that is, the capacity of the gas to absorb energy without having the temperature go up. The originality of Waterston's ideas should have caused notice . . . but he didn't fare much better than Herapath. In some ways he fared worse.

John James Waterston (1811–1883) was another product of the British Isles and as the son of a wax manufacturer, he, too, had his first contact with chemistry though an industrial process.[5] Waterston wrote his first paper at the age of nineteen, and in it he sought to explain gravity without resorting to Newton's mysterious action at a distance. To do this he invoked collisions between particles as a means of transferring influence. At twenty-one, needing employment, Waterston went to work for the railroad as a surveyor. He found the work too intrusive on his time for scientific research, so he applied for employment with the East India Company, which meant he would have access to Grant College in Bombay and more leisure time to spend there. In 1843 he sent home a manuscript for a short book that was published anonymously. In the book he tackled several topics, including the physiology of the nervous system and a kinetic theory of gases that contained one of the first concrete statements of the equipartition theorem.

The equipartition theorem, or the theory of equal partition of energy, says that energy will be equally distributed in all modes. For a gas made of individual atoms, this means that the atoms are as likely to be flying up as down, right as left, and all directions of the compass as well. This is because atoms colliding in a gas can be assumed to behave like hard billiard balls, colliding in perfectly elastic collisions, that is, not sticking together. But recall that molecules are more like jellyfish. When one billiard ball strikes another, the energy from the first will be transferred into translational energy for the second, and the object ball moves off straight and true, but the situation is different for two colliding jellyfish. The object jellyfish will absorb some of the energy in vibrational and rotational modes as well as translation. The equipartition theorem says that energy will be equally distributed in all the modes, that is, all the energy won't just suddenly and spontaneously concentrate in one particular vibrational mode or suddenly all concentrate in translation motion, abruptly sending the jellyfish flying across the room. However, the name of the book, *Thoughts on the Mental Functions,* did little to alert potential readers to the fact that it contained a kinetic theory of gases, let alone an equipartition theorem, so Waterston's theory went neglected.[6]

In December 1845 Waterston sought to rectify this situation by submitting to the Royal Society a paper entitled "On the physics of media that are composed of free and elastic molecules in a state of motion."[7] As with Herapath's paper, the referees rejected it, one of whom went so far as to call it "nonsense,"[8] but unlike Herapath, its title was read into the meeting notes, and in that act, made it unavailable for submission to another outlet.

To understand the situation, it must be recognized that today it is still considered highly unethical to attempt to publish the same material twice. So the policy of the Royal Society was that once a paper was read into the meeting notes it became the property of the society. In Herapath's case, he sensed the winds were against him, and elected not to have his paper read, which allowed him to ask for it back so he could publish it elsewhere. But Waterston was in India and did not, or could not, exercise this option. The paper remained inaccessible in the archives of the Royal Society until Lord Rayleigh found it in 1891, after Waterston's death, and saw that it was published.

Waterston's idea of using the measured ratio of the heat capacity at constant pressure to the heat capacity at constant volume to compare with the predictions of the kinetic theory of gases turned out to be enormously useful and would eventually lead to necessary refinements in the theory. To explain why the heat capacities should be different for a gas at constant pressure and the same gas at constant volume, consider two swarms of gnats. One swarm is contained in a cylinder with rigid walls and another swarm in a cylinder with one movable wall, such as a piston. If energy is added to these systems, the temperature behaves differently. When energy is added to the swarm in the cylinder with the rigid walls, the temperature goes up significantly because energy is going into increasing the translational energy of the gnats. For the other system, however, the one with the one movable wall, some of the gnats will strike this wall with their additional energy, and the wall will move, absorbing the energy. So the gnats in the cylinder with the piston will have to take in more energy for the temperature to go up by the same amount. The system with the movable wall will have a higher heat capacity, and the kinetic theory of gases should predict exactly how much. But for reasons known only to gnats, the situation is complicated by quantum mechanics (which we will consider later), and though the numbers Waterston calculated agreed with the experimental results of others, which gave him confidence in his theory, they were unfortunately based on an arithmetic error. Had he not made the error, he would have gotten different results, which would have been more discouraging.

Waterston returned to Scotland in 1857 and with his accumulated savings attempted to establish himself as a scientist, but met with little success. He spent the rest of his days in retirement, visiting his nephew, and playing billiards and chess. By all accounts he was resigned to his unusual and out-of-step life, though reportedly used "unparliamentary language" when speaking of new scientific advancements.[9] Then one day he took a walk on a breakwater and did not return. It is doubtful that Waterston was suicidal, but one wonders if he resisted the waves that carried him away.[10]

But Waterston and Herapath might have taken some consolation in the fact that even Joule (who may have used Herapath's theory as a starting point) attempted the same type of analysis and basically met with the same reception. In fact, the kinetic theory of gases did not really attract attention until 1856 when Karl Krönig published a short paper on kinetic theory that was essentially an encapsulation of the work of Bernoulli, Herapath, Waterston, and Joule. But it had the advantage that the time was right. As may be recalled from our discussion of thermodynamics, Helmholtz, by then, had demonstrated the conservation of energy. Joule had demonstrated the equivalence of heat and mechanical motion. Clausius needed to get a better handle on entropy. The kinetic theory came into the hands of the master.

Rudolf Clausius was aware that the theory didn't match all the data, so he used a tactic that has often proved fruitful in chemical physics: He assumed, for the time, the missing part was simply a proportionality constant. In other words, he called it potatoes and moved on. In an 1857 publication he presented a concise kinetic theory of gases and clearly introduced the concept of *mean-free path*, that is, the average length of the path that a particle travels before it collides with another particle. The idea of a gas particle's path being short before it changes direction satisfactorily explained why gases and odors don't instantaneously diffuse from one side of the room to the other. Clausius also suggested that there must be a distribution of velocities rather that just one velocity for all gas particles, which explains why evaporation takes time: A pot of water isn't liquid one second and a cloud of steam the next. However, Clausius did not take this important idea any further. It was left to James Clerk Maxwell, as we shall see next, to derive an expression for the velocity distribution of a gas. Clausius also left Maxwell and colleagues with the challenge of assigning meaning to his nebulous proportionality constants.

Now the stage is set for Ludwig Boltzmann, that is, Boltzmann and a couple of friends—Maxwell and Maxwell's demon—and he'd have a devil of a time taming them. As he would summarize, "I am conscious of being only an individual struggling weakly against the stream of time. But it still remains in my power to contribute in such a way that, when the theory of gases is again revived, not too much will have to be rediscovered."[11] Bernoulli's paper had been republished in 1859, before the death of Herapath or Waterston, and Lord Rayleigh would uncover Waterston's paper in the prime of Boltzmann's career, so Boltzmann was painfully aware of the fate of Herapath and Waterston, and hoped to avoid it. In the end, he not only escaped their obscurity, he realized their ambition and his own.

Chapter 14

Statistical Mechanics

I have never seen an astrologer who was lucky at gambling . . .
Girolamo Cardano, circa 1550

The advent of statistical mechanics as a predictive tool in physical science is a major turning point in the history of physical chemistry and indeed in the history of science—but it was not an easy transition. Science is the search for system, and nothing pleases a scientist more than to find a key that fits. So when Newton's mechanics explained planetary orbits, cannonball trajectories, and, via the kinetic theory of gases, the behavior of air, no one wanted to abandon Newton. It fell to Boltzmann to remind his brethren that they had more work to do.

Not only did Boltzmann have the dubious distinction of issuing this wake-up call, he had to deliver a bitter pill with it: Not only would science have to abandon the security of Newton, it would have to relinquish his certitude, too. With the arrival of statistical mechanics we see Newton's universe of constants, absolutes, and certainties yielding to maybes, most likelies, and probabilities. But in some ways it was a return to an old idea. Who knew better than the medieval mathematicians that the future is a gamble?

To begin, we revisit an old friend, the mathematician Girolamo Cardano, and hear a tale he relates in his *Book on Games of Chance*, written around 1550.[1] He conveys his own story of being twenty-five and experiencing some sexual dysfunction. He met a man who said he knew a beautiful prostitute who would soon set Cardano to rights. It turned out to be a lure to a dice game instead, and once baited, Cardano found he lost all his money and even some of his clothes. Cardano, as he tells it, went home and thought over the situation. In his ponderings, he realized that in the game he had seen certain numbers show up on the dice more often than others. He calculated that seven was the most likely number to show up because there were six ways of achieving it on two dice and that two and twelve were the least likely because there is only one roll each that would produce these numbers. He then decided to go back to the dice game with this knowledge, but took the precaution to tell his servant to interrupt him as soon as he had won back his clothes and money. He then bet only when the odds were in his favor and soon won back what he had lost. He noted that a fellow gambler commented that Cardano acted as though advised by a demon, but the loser accepted his losses graciously. Cardano adds that this is well because he knew of a man who had blasphemed after a gambling loss and had been arrested and condemned to hang. "Christians," he comments, "tolerate gambling . . . but they do not allow cursing."[2]

Although Cardano's application may have been less than scientific, Boltzmann and physical chemistry owe much to the gamblers and gamers of the backstreets of the Renaissance and to the later card salons of Europe. In the 1600s, Blaise Pascal, a moralist who was also quite interested in mathematics, shared a long coach trip with a free-living nobleman by the name of Chevalier de Méré.[3] To pass the time, they discussed an old gambling problem: how to split the pot in a dice game that had to be discontinued before the final die was cast. After the coach ride, Pascal communicated the problem to a fellow mathematician and together they agreed that the pot should be split according to each player's probability of throwing a winning number, had there been time enough for them to take a turn. Their efforts at calculating probabilities were soon generalized to any situation with multiple possible outcomes. In the 1700s Simon Laplace, a younger and politically more astute fellow bourgeoisie and fellow scientist of Lavoisier, built on this foundation. It is sometimes said that Laplace survived the revolution because the government needed him to calculate trajectories for the artillery, but he also had the good sense to test the political winds and align himself with the powers that be. An extremely talented physicist and mathematician, he designed a probability theory that allowed him to extract information from astronomical observations that were limited in their precision only by the precision of the instruments. The data he gleaned with his theory for handling error allowed him to take Newton's calculations a few steps further and add some corrections to make them more accurate. Newton himself had credited a nudge from God to correct for the occasional discrepancies in his orbital theory. When Napoleon asked Laplace why he did not evoke the Creator in his system, he said "I have no need of that hypothesis."[4] Though this statement may or may not have been atheistic, it demonstrates that Laplace believed he had perfected his methods to the point where he did not need to invoke divine intervention. With Laplace, statistics had found their place in physics.

Statistical methods were imported to England in 1850 through John Herschel (1792–1871), son of William Herschel, the astronomer who discovered Uranus.[5] John Herschel learned mathematics, chemistry, and physics at home from his parents and his aunt, Caroline Herschel, who was a noted astronomer in her own right.[6] Ten years later, Scottish mathematician and physicist James Clerk Maxwell applied statistical methods to the kinetic theory of gases.

One cannot begin a discussion of James Clerk Maxwell (1831–1879) without first noting that he contributed to several important areas of physics, all before dying at the age of forty-eight. Like Lavoisier, he had his finger in many pies. Here, however, we first encounter him in the context of his contribution to the statistical treatment of the kinetic theory of gases.[7]

Maxwell's parents also planned to educate him at home, but in a scenario all too common in those days, Maxwell's mother died while he was still quite young, so he ended up attending Edinburgh Academy. At the academy he was dubbed "Dafty" because he was introverted and spent his time reading, drawing diagrams, and making mechanical models, but when he started winning prizes for scholarship, math, and poetry, the nickname vanished. He submitted his first paper to the Royal Society at the age of fourteen, and though it was only a revision of an older problem in Cartesian geometry, it was impressive for a fourteen-year-old. Maxwell earned an academic post by 1857 and when one contest was announced for the best explanation of Saturn's rings, he decided to try his hand. His solution was to show that the rings could be stable if they were made up of small particles rather than one continuous ring. He defended his thesis sufficiently to win the prize and twentieth-century observations of the rings of Saturn proved him right. Later, the problem of a collection of colliding small particles (the kinetic theory's swarm of gnats) became a constant theme in Maxwell's work.

Another constant in his life after June 1859 was Katherine Mary Dewar Maxwell, his wife. She helped him extensively in his work, as evidenced by the contents of a postcard from Maxwell to his old friend and fellow student at Edinburgh Academy, Peter Tait: "My better 1/2, who did all the real work of the kinetic theory is at present engaged in other researches. When she is done I will let you know her answer to your inquiry [about experimental data]."[8]

The data they were collecting was to test Maxwell's treatment of the kinetic theory of gases. Though it had been proposed that after enough collisions the molecules would have all exchanged energy so much that they would all share the same velocity, Rudolf Clausius had seen that this did not match experience (a glob of gas does not move across a room in lock step but drifts with reaching fingers). Building on the work of Clausius and using Laplace's error function (the famous or infamous "bell curve"), he showed that many velocities were possible in a gas at equilibrium and the probability for a particular velocity depends on the curve and the magnitude of the velocity; thus the resulting distribution of velocities is a skewed curve that has a few particles at low velocity and a few at high velocity, and even some at very high velocity, but most in the middle. However, because it was not an exact bell curve, the average velocity is not exactly in the middle. The experimental velocity distribution of a gas can be measured by blocking the path of a gas with two rotating, slotted disks that allow a blob of gas to pass only when it has the right speed to make it through both slots. An analogy would be traffic lights timed so that people going exactly the right speed traverse a series of intersections without stopping. Conversely the average speed distribution of cars on that road could be measured by varying the timing on the light to see how many cars made it through at various settings. To measure the average speed distribution of a gas, the timing of the slots is varied and the amount of the gas that passes through is measured. The average predicted by Maxwell's skewed curve agreed with the experimental average velocity.

Maxwell then decided to re-examine the kinetic theory of gases. He knew of its failure to predict the correct ratio of heat capacities for the gases studied at that time, and he hoped to find the theory's flaw. As we mentioned previously, John Waterston's agreement with experiment was a happy accident and actually an arithmetic error. Maxwell was less prone to such errors (though not immune) and he knew he faced a real discrepancy between experiment and theory.

In explaining the discrepancy, we must realize that it was known then that the common gases—oxygen, nitrogen, hydrogen—were diatomics; that is, they consisted of two atoms per molecule. A gas made up of diatomics can absorb more heat before its temperature goes up; that is, it has more ways that it can move (more like a jellyfish than a billiard ball). It turns out that the ratio of heat capacities is predicted to be 1:3 to account for all those motions. But in actuality the heat capacities kept obstinately measuring 1:4. They had to admit the jellyfish were a bit tougher than predicted, which meant failure for the kinetic theory when it came to diatomics.

Though Maxwell never solved the heat capacity ratio problem (the full solution would have to await quantum mechanics), he found another result that pleased him so much that he was able to overlook the heat capacity enigma and support the kinetic theory of gases.

What Maxwell found was that the kinetic theory seemed to imply that the viscosity coefficient for a gas should be independent of pressure and go up with the square root of the temperature. In other words, if you squeezed the gas into a smaller volume, it did not get more viscous, and if you heated it up, it did. In our normal day-to-day experience, most of us have come to equate more dense with more viscous and higher pressures of gases mean a denser gas. We certainly equate a lower temperature with a higher viscosity (for example the expression "slower than molasses in January"). While Maxwell's result seemed as counterintuitive to him as it does to most people now, he was unable to find his error, so he courageously went back and revisited the experimental data for gases. He and his wife were able to show experimentally that rarefied gas viscosities were indeed independent of pressure and did go up as the temperature went up, and when Maxwell was able to rationalize it theoretically in terms of the kinetic theory of gases, he became a believer.

To understand why this might be so, it is a good idea to look at what viscosity is. Viscosity is the resistance to flow in a given direction. A more viscous liquid (molasses) will resist being poured down while a less viscous liquid (water) pours more readily. In a liquid, the molecules are relatively close together and the attractions between the sticky molecules make them resist flow. When liquid is heated up, the molecules move around faster, the sticky attractions become less of a factor, and they do not resist flow as much. In a gas, however, the molecules are already so far apart that the attractions between the molecules aren't as important (just as separating two magnets diminishes the force felt between them). So the viscosity of a gas must be modeled by something nonsticky: for instance, cows being herded down a cattle chute.

In a cattle chute, the cows in the center tend to be moving faster because the ones on the edges are bumping into the fence, which slows them down. Collisions between the fast cows and the slow cows speed up the slow cows and slow down the fast cows. At a higher density of cows (that is, a higher pressure) there are more collisions, but of both types: More fast cows are being slowed down by collisions with slow cows and more slow cows are being speeded up by collisions with fast cows. Of course there are cow psychology and ethical questions that we are ignoring in this analogy, but atoms have no psychology, so pressure has no net effect on gas viscosity.

Now let's look at the increase in temperature. Viscosity is resistance to movement in a given direction. When the gas is heated, more energy is added to the whole gas, but not just in one direction. Borrowing our previous analogy, the herd is spooked, but not from just behind but from all directions at once. This will throw the herd into a random panic, which is not a bad analogy for what happens to the gas. Gas particles are now moving faster in all directions, and in fact interfere with movement in the selected direction, so viscosity increases.

Having put the viscosity problem to bed, Maxwell was tempted to tackle the more difficult question of a molecular basis for entropy. He had an inkling, however, as to how the explanation would have to go. Maxwell felt that the explanation would lie in the statistical treatment of gases, based on the known laws of physics, not on the discovery of some new physical law. Maxwell saw the situation not as a rule of nature but rather as a result of our clumsy attempts to model nature. In fact he could imagine a way to circumvent entropy if only the human frame were finer: A diminutive demon could sit at the boundary of a hot material and a cold material, operating a shutter, and defeat entropy by allowing only the cold atoms to move to the hot side and stop the hot atoms from going to the cold side. Maxwell's demon first appeared in a letter to Peter Guthrie Tait on December 11, 1867.[9] But though he defined the problem, Maxwell did not devise the equations. Ludwig Boltzmann (1844–1906), a university professor thirteen years Maxwell's junior, would take the heat for defining the devil.[10]

Boltzmann's father was a bureaucrat in the office of taxation in Vienna. The son, a perceptive student, received his doctorate in physics from the University of Vienna at the age of twenty-two. In the process of securing an academic appointment, he worked on the kinetic theory of gases and extended Maxwell's distribution of gas velocities to a distribution of energies and in doing so defined the proportionality constant between temperature and energy that has since been named Boltzmann's constant. Boltzmann's constant has

Fig. 15. Ludwig Boltzmann. AIP Emilio Segrè Visual Archives, Segrè Collection.

quite a respectable place in all of physical chemistry, but for a time, it went unheralded by everyone, including Boltzmann. However, the theory attracted enough attention so that the twenty-five-year-old was offered a university position while other, older candidates continued their tutoring and waiting. Boltzmann then, too, turned his attention to entropy, though he had to try several approaches before he settled on his solution. When he did, he inaugurated a new paradigm in science.

Building on Maxwell's distribution of velocities, Boltzmann showed that a gas in any arbitrary initial state will continue changing its velocity distribution through collision until it reaches the skewed-bell-curve Maxwellian distribution; at this point it will continue to experience collisions, but cease to evolve. When Boltzmann identified these velocity changes with entropy, he was able to recast his result and say that a system will tend to increase in entropy until it reaches its equilibrium state. In other words, an ordered deck of cards becomes disordered on shuffling, but on continued shuffling it eventually reaches its most chaotic state and though shuffled forever, will not become more disordered. Boltzmann's theory was a major triumph for statistical mechanics (though not yet named "statistical mechanics"), but in its day it was received with skepticism, to say the least.

One objection was the so-called reversibility paradox.[11] In mechanical systems, it is a given that any forward process can be reversed. If a ball rolling downhill produces energy, exactly the same amount of energy is required to push it back up the hill.[12] But entropy is unidirectional. Systems became more chaotic, not less. So, went the conclusion, entropy could not be based on a simple mechanical process such as atoms colliding with each other, as Boltzmann had presumed. If so, why shouldn't systems just as spontaneously order as disorder? Reinforcing this objection was the idea of recurrence, that a mechanical system forced to move in a finite volume would eventually return to its initial conditions (a marble randomly bouncing around in a box will eventually hit its beginning point). Boltzmann knew intuitively that he was not contradicting either of these principles, it was just that the probability of their occurrence was extremely small.

To understand Boltzmann's entropy, let's revisit our card game. Assuming we are playing with a full deck (let's hope we are), that means we have a deck of fifty-two cards that can be divided up into four suits of thirteen cards each. If we start with a well-shuffled, chaotic deck, and deal out all the cards to four players, the odds of dealing one particular ordered hand, for instance a thirteen-card hand that is all spades, is on the order of 640 billion to one.[13] And this is only one hand. To have all the hands so ordered would be against astronomical odds. But if this hand is considered ordered, and every other hand is considered disordered, then there are 640 billion hands that qualify for disorder. This was Boltzmann's point: Equilibrium is not one configuration, but many, many configurations. So the odds for disorder were just so much more probable that one would not witness a spontaneous return to order in one lifetime or many lifetimes or a universe of lifetimes. In 1877, Boltzmann derived his famous relationship that entropy, S, is equal to k (Boltzmann's constant) times the logarithm of the number of ways a particular configuration can be achieved, $S = k \log W$. This equation is engraved on his tombstone.

Within ten years of Boltzmann's death, new information and new philosophies would vindicate his theories and validate his results. Even as early as 1875 information began to gather on his side. When the ratio of heat capacities for monatomic gases based on billiard balls were calculated, no monatomic gases were known. In 1875, however, August Adolph Kundt (1839–1894) and Emil Gabriel Warberg (1846–1931) found one.[14] They had been working on measuring the heat capacity ratios of gases by measuring the speed of sound. Sound is a compression wave that propagates through a gas by rapidly squeezing the gas and then allowing it to expand. The rate at which the gas compresses and expands depends on its compressibility, which in turn is related to its heat capacity. By setting up a standing wave of sound (which can be obtained by blowing across the open mouth of a narrow-necked bottle, though they did it by stroking a glass rod) they could track the crests and valleys of sound in a finely divided dust. From these dust figures they calculated the speed of sound and the heat capacity ratios for many gases.

What they found was that mercury vapor has a heat capacity ratio that is in perfect agreement with that predicted by the kinetic theory of gases for a monatomic gas. There was still that troublesome business about the heat capacity ratio for diatomic gases, but Boltzmann arrived at what he found to be a convincing explanation in 1877. Suppose, he said, that a rotating diatomic is like a pencil rotating on a table. Then there are three mutually exclusive axes about which it can rotate: one rotation flat on the table, one rotation end over end as the pencil does cartwheels down the table, and one in which the pencil just rolls on the table. He then proposed that this last rotational mode would be unaffected by collision: A direct hit on one end of the pencil could set it spinning flat or cartwheeling, but a *direct* hit to the center of the pencil should just push it forward, not set it rolling (to get rolling it would have to be hit off center, that is, slightly high or slightly low). By subtracting out this mode of motion as a place to put heat, the ratio of constant pressure heat capacity to constant volume heat capacity became 1:4 for diatomics, in excellent agreement with the measured heat capacities.

Maxwell, weakened by the abdominal cancer that was to kill him, did not accept Bolzmann's explanation for the anomalous heat capacity ratio, though many other physicists did.[15] As it turned out, the true explanation for the missing mode would come only from quantum mechanics, but there were plenty of other reasons to support Boltzmann's statistical view of entropy. One Boltzmann supporter in particular would not only accept Boltzmann's approach but reinforce it, extend it, and name it the science of statistical mechanics: U.S. citizen Willard Gibbs. In 1902 Gibbs would bring statistical mechanics to chemistry in his book *Elementary Principles in Statistical Mechanics*. But Boltzmann himself would work no more on his theory. In 1906, while on holiday and while his wife and daughter were out swimming, he hanged himself from a balcony window.

Chapter 15

Thermochemistry

A mathematician may say anything he pleases, but a physicist must be at least partially sane.

J. Willard Gibbs, circa 1870

In our discussion of thermodynamics and statistical mechanics thus far, there has been one notable omission: chemistry. We've discussed change, but only physical change such as water expanding into steam, but remaining H_2O. Chemistry is about another kind of change—material change—salts into semiconductors, dough into bread, petroleum into plastic, and plants into pills. Chemistry is what happens when our swarm of gnats evolves into a swarm of bees. In fact, not just a swarm of bees, but a conglomeration of bugs: some bees, some beetles, and some gnats left over, too. Chemical reactions do not neatly change all of one material into all of another, but instead result in an equilibrium mixture of both beginning reactants and the final products. It was this convoluted problem that was tackled by the steady hand of Willard Gibbs.

To understand the role of equilibrium in chemical reactions, consider why cleaning a floor with dirty water won't work. In mopping a floor, the dirt on the floor and the clean, soapy water can be seen as the beginning reactants. The dirty water with dissolved dirt is the final product. When the clean, soapy water is first spread over the floor, dirt dissolves in it, but eventually the water reaches its equilibrium dirt load and the floor won't become cleaner until the dirty water is mopped up and replaced with fresh. At equilibrium, some of the dirt is dissolved in the water, but some is still left on the floor. It would take an infinite number of swabbings to produce a floor that is perfectly clean.

Before the advent of thermodynamics, chemists were aware of this tendency in chemical reactions to proceed just so far, and no further, and had identified at least one factor that governed the final amount of product: the mass of the reactants. In other words, the greater the amount of dirt on the floor, the greater the dirt load of the water with the first mopping. In 1864, Peter Waage and Cato Maxmilian Guldberg, both of Norway, quantified this observation by careful experimentation and showed that at a particular temperature, the equilibrium ratio of all the mass of product produced to all the mass of left-over reactant is a constant. In other words, if one had an equilibrium mixture of products and reactants and added a bit more reactant, then more product would be produced until equilibrium was restored, or visa versa. The usual visual analogy offered is that of a board balanced on a pivot point: If weight is added to one side of the board, then weight must be added to the other side of the board to restore equilibrium. This quantified observation of Guldberg and Waage is called the law of mass action.

The law of mass action explains why we are able to breathe. Free hemoglobin and oxygen are the reactants while the bound oxygen-hemoglobin complex is the product. In the lungs, the excess of oxygen in the oxygen-rich air forces the formation of more oxygen-hemoglobin complex. The blood containing this complex then travels to the cells where oxygen has been depleted. The situation then shifts in favor of free hemoglobin and oxygen, and oxygen flows into the oxygen-poor cells.[1]

But there is more than mass involved in equilibrium, there is also energy. The fact that energy is an important ingredient in chemical reactions can be seen in cooking: Mixing the ingredients for a cake together and letting them sit on the table will not create a cake. Energy has to be added along with the eggs, flour, and sugar. The mixture has to be put in the oven.

The fact that energy is a product of some chemical reactions can be attested to by anyone who has placed themselves close to a combustion reaction, that is, close to a fire. But the role of energy in equilibrium was confusing to early investigators. It seemed reasonable that spontaneous reactions should all give off energy. If the reactants were attracted to each other, then coming together should take them to a lower energy state. The situation can be modeled by magnets: It requires energy to keep them apart; it is a lower energy situation to allow them to come together. However, several spontaneous reactions were known that required energy rather than produced energy. Instant cold packs used nowadays by athletes are made of two chemicals that become cooler when they mix. When the chemicals mix, the reaction consumes energy, absorbing it from the skin and surrounding air. The fact that a chemical reaction should consume energy seemed counterintuitive. It was not until Clausius introduced the concept of entropy that researchers began to see how spontaneous energy-consuming reactions could occur.

We now know that the change in energy over the course of a reaction is the difference between the energy required to break the chemical bonds of the reactants and of the energy gained when chemical bonds re-form in the products. An analogy might be a trip to Las Vegas. The total cost of the trip is the difference between the money spent on the trip and the money won at gambling. Sometimes there is a net loss, sometimes a net gain. If it costs more energy to break bonds than is gained back by forming products, then the reaction requires energy. If more energy is recovered when the products are formed, then the reaction gives off energy. Reactions that cost energy overall can occur spontaneously if the reaction increases the entropy. Equilibrium is the point at which there is balance between the energy costs of the reaction and the entropy gains.

We can explain the role of entropy in equilibrium in terms of two rival movie theaters, on opposite sides of town, both showing the same movie. One theater decides to attract customers by providing ample seating; that is, there is plenty of "elbow room" or room to spread out. Entropy can be thought of as the tendency to spread. To paraphrase Maxwell, entropy explains why a tumbler of water thrown in the ocean can never be retrieved.[2] The wish to spread out is a human desire, too. The other theater decides to attract customers by selling cheaper tickets, a cost factor that for us represents energy. Therefore there are two factors that result in the distribution of people between the two theaters: People will crowd into the cheaper theater until they start to feel claustrophobic, then they will shift their patronage to the more expensive theater for the sake of spreading out. Once some of the people have crossed town to the more expensive theater, the crowding is alleviated in the cheaper house, and the migration stops. When the lights dim and the movies start, there will be a fairly predictable ratio of the number of people in the cheaper theater to people in the more expensive one. In chemical reactions, if enough entropy is gained by reactants turning to products, then the reaction may still be spontaneous, even though it costs energy.

Now let's take the analogy a bit further. The law of mass action says if more reactant is added, then the reaction will produce more product until equilibrium is restored. In our theater situation, this might be an influx of students back to their two-theater town during summer break. Of course the majority of the new patrons will flood into the cheap theater, but as soon as they feel their space invaded they may be willing to pay the higher price to spread out. This shift, however, results in more-than-usual crowding at the higher priced theater, so eventually people are less willing to spend more money for a small gain in privacy. At equilibrium, there will be more people on average in each theater, but the ratio of people in the more expensive theater to the people in the cheap seats will be about the same.

Now let's take the analogy even further. If the higher priced theater decides to take advantage of the new situation and raise its price (the energy cost for the reaction is greater) the equilibrium ratio will shift in favor of the lower priced theater; people are willing to pay only so much for comfort. So in theaters, as well as chemical reactions, energy and entropy are both factors in determining the final ratio of populations, but they play against each other. To understand the fine balance of equilibrium requires an awareness of the principles of both physics and chemistry, which is a true test of physical chemistry. With quiet conviction, Willard Gibbs undertook the problem of untying this Gordian knot. But then Willard Gibbs came from a family used to difficult dilemmas and who had developed a patient, practical approach to their untangling.

Josiah Willard Gibbs (1839–1903), whom biographers designate Willard Gibbs to distinguish him from his father, also Josiah Gibbs, was born New Haven, Connecticut, when it was a bustling, brawling, bawdy seaport in a country on the verge of civil war.[3] But the Gibbs family managed to hold its dignity in the midst of this savagery. When the African captives who rebelled and took over the slave ship the *Amistad* were seized and jailed in New Haven, Josiah Gibbs, Willard Gibbs's father, professor of theology and sacred literature at Yale, helped them in a unique, quiet way. He learned the language of the Africans and by communicating with them established their origin and legal status.[4]

The son, Willard Gibbs, must have inherited this penchant for quiet conviction and practicality from his father. As a child, Gibbs was too reserved and too academically inclined to be socially acceptable. He was teased by his fellow students and generally treated as an annoyance. However, he does not appear to have expended much concern over this, instead enjoying the company of his family and his studies. He entered Yale, as expected, and continued there until he had received a Ph.D. in engineering, the first awarded at Yale and one of the first in the United States.[5] He remained at Yale as a tutor, not unduly concerned with money because his parents had left him and his sisters a comfortable inheritance. He worked on and received a patent for an improved railroad brake, then decided to go to Europe and study for three years, a common plan in those days for aspirants to a sound scientific education.

He was accompanied on this tour by his two sisters and when they returned they all moved back into the family home, though one of the sisters eventually married and had a family. Two years after his return from Europe he received an appointment as professor of mathematical physics at Yale. This appointment may have seemed surprising because it was made before the thirty-two-year-old Gibbs had published any work (though he did have a patent). However there was probably not a great clamor for positions at the provincial backwater of New Haven and Gibbs himself went without salary for

Fig. 16. Josiah Willard Gibbs. Edgar Fahs Smith Collection, University of Pennsylvania Library.

the first nine years. Once situated, however, Gibbs tackled the problem of chemical equilibrium.

Gibbs began by clearly defining entropy, a necessary first step. Peter Tait had confused its meaning in his book on thermodynamics and his friend Maxwell had propagated the error. Gibbs, not counting on his papers being discovered in the *Transactions of the Connecticut Academy* in which they had been published, sent copies to important European chemists and physicists, including Maxwell. Maxwell, on reading and working his way through Gibbs's complex mathematical analysis, saw his own error and enthusiastically embraced Gibbs's approach.

Gibbs then proceeded to describe a new thermodynamic quantity that would relate the entropy and the energy cost or gain of the reaction to the equilibrium distribution of products and reactants. The quantity he came up with is today called "Gibbs's free energy," though Gibbs himself did not use that term (it is considered poor form to name things after yourself). What Gibbs managed to define in free energy is the pivot point for the balance board of chemical equilibrium. Gibbs showed that by calculating the change in energy and the change in entropy in going from reactants to products, and balancing them against one another, one could calculate how much energy was still "free" and this would predict the direction of the reaction. For instance, if there were no free energy, then the reaction would be at equilibrium and the reaction mixture would not change. If there were some amount of net free energy, then the reaction would either produce more product or more reactant until it reached equilibrium. In the process of approaching equilibrium, like water falling from high ground to low, a chemical system can also do work; therefore free energy can be thought of as the energy free to do work.

In our theater analogy, calculating the free energy would be tantamount to using data about the number of seats and the cost per seat to predict which theater would attract the most people. A successful marketing consultant who could tell a theater owner exactly how to build a theater and how much to charge would be very valuable indeed. In free energy, Gibbs gave chemists a valuable tool for controlling chemical reactions.

How important was this? Perhaps the importance is best conveyed by the story of a French chemist of the day, Henry Le Chatelier, who was engaged in an industrial problem. His task was to make ammonia from nitrogen and hydrogen. Ammonia is a fairly simple molecule, built from one nitrogen atom and three hydrogen atoms, and nitrogen and hydrogen are readily available reactants. But simply mixing nitrogen and hydrogen together doesn't produce a large quantity of ammonia, so it was a problem in equilibrium—and one that needed to be solved. Ammonia is used to make nitrates and nitrates are used to make gunpowder and Europe at the turn of the nineteenth century was on the brink of war.

Le Chatelier needed to find a way to tweak mixtures of nitrogen and hydrogen to make them produce more ammonia. But which factor should he change? Temperature? Pressure? Amount of reactants? If he tried them all individually, it would take him years, and he didn't have years. He tried to force the reaction to the product side by turning up the pressure and the temperature at the same time; unfortunately some oxygen accidentally contaminated his high-pressure chamber. When the dust had settled, some pieces of the chamber were found in the ceiling and some embedded in the floor. Le Chatelier gave up that approach, though an analysis of Gibbs's free energy would have shown him he was on the right track. Within five years, the German Fritz Haber found the magic combination.[6]

When Le Chatelier learned of Gibbs's analysis of equilibrium, he was understandably enthusiastic. Gibbs's equations showed how a calculation could be done to predict how temperature and pressure would affect an equilibrium. In a matter of hours reactions could be designed on paper that would otherwise take months or years of laboratory trial-and-error. It was a chemist's promised land . . . but with one catch. Not all the data to plug into Gibbs's equations was available yet.

Progress was being made by measuring energy produced or consumed by reactions. Gibbs himself had shown a way to experimentally measure some free energies via electrical potentials. Gibbs had also worked on Boltzmann's and Maxwell's idea of the distribution of energy in the various motions of molecules. He even christened the science by calling his book on the subject—his last book—*Elementary Principles in Statistical Mechanics*. He had devised computational methods that would allow the calculation of the total energy of a system from the energy of the molecules . . . but therein lay the other rub. Physicists knew how energy should be absorbed by masses on a spring. They could model the energy distribution of a pendulum and collection of cannonballs. It seemed only reasonable that molecules made from atoms should behave in similar ways. But the pesky molecules just kept refusing to fit their mechanical models. Assuming molecules behaved as vibrating and rotating masses on springs just wasn't providing the right total energies. To make matters worse, there appeared to be some question as to whether atoms were even the billiard balls of Dalton's vision. Studies of the interaction of light with atoms seemed to show that they were absorbing energy from the light in what looked like vibrational motions. In response to this conflicting information, theorists once again began casting about for different fundamental models of matter. Was it time to ditch Dalton? Was it time to scratch the whole system and start fresh? One thing was certain, it was time to answer this question: Are there atoms or not?

It was a physical chemist who stepped forward to take on the challenge: Jean Perrin.

Chapter 16

Atoms or Not?

. . . these articles simply serve to reveal an internal agitation of the fluid
. . . much as a cork follows . . . the movements of the waves of the sea.
Jean Perrin, 1909

Jean Perrin is sometimes a forgotten hero. A survey of modern text-books reveals it is quite possible to write a text on physical chemistry without mentioning Perrin at all. Perhaps this neglect is because he did not invent an equation that is in daily use like Gibbs or find a constant that would bear his name like Boltzmann. Perhaps it is because Perrin's intricate and exacting experiments, once done, did not have to be repeated, and do not have to be redone each time a new atom is discovered. But we need to remember Perrin. He reminds us that we must always question the basis of our assumptions. Even the atomic model had to be proven and could not be accepted on intellectual appeal alone. Fortunately he was not so neglected in his own day: This physical chemist won the Nobel Prize for physics.

When Perrin set out to prove the existence of atoms, he knew what he was up against. Josef Loschmidt (1821–1895) had made an estimate of the diameter of an air molecule (nitrogen or oxygen), based on the kinetic molecular theory and known pressures.[1] His estimate was about a billionth of a meter, which was still about four times too big. To imagine how infinitesimal this is, an atom would have to be magnified a million times to be seen with the naked eye. Loschmidt also estimated another quantity of importance to all physical chemists and indeed all chemists: Avogadro's number.

186

The number is usually called Avogadro's number to honor the Italian lawyer turned physicist, Amedo Avogadro (1776–1856), though Avogadro did not actually calculate Avogadro's number.[2] It was Perrin, not Loschmidt, who named this quantity Avogadro's number (some still call it Loschmidt's number), but whatever it was called, all physicists and chemists of the day were interested in it. Simply put, Avogadro's number is the number of atoms in a handful of sand or in a big balloon filled with air. It is the number that allows for the calculation of the mass of one atom. No scale can weigh one atom, but if the mass of a handful of atoms is known, and the number of atoms in the handful, then the mass of an individual atom can be calculated by dividing the total mass by the total number. Dalton had suggested a method for finding the *relative* masses of some elements, but he had not shown how to arrive at the absolute mass. Loschmidt and others realized that if they could find the number of particles that made up a given reference quantity, then by weighing this quantity and dividing by the number of particles, one could find the absolute mass. A new method to weigh gnats.

The reference quantity they decided on is called a *mole*. One of the many uses of the word *mole*—from a blind, furry animal to a skin imperfection—is to connote a large sample of something.[3] The word was first used in the context of chemicals by another champion of physical chemistry, Wilhelm Ostwald (1853–1932). But when Ostwald used the word *mole*, he was not thinking in terms of molecules. Ostwald was one of the detractors of the atomic theory and one of the people Perrin had to convince.[4]

When Loschmidt and others made the initial estimates of the number of molecules in a mole, the answer was astounding. A mole, which is about a handful of solid or about twenty-two liters of gas under normal conditions of temperature and pressure (a little more than a bucket and a little less than a bushel), is some 600,000 billion billion molecules. That's a lot of gnats.

Many analogies have been drawn to try to convey the enormity of this number, and considering these is instructive. For instance, it has been calculated that one mole of marbles would cover the land area of the United States to a depth of 4 meters (about the height of a one-story house not counting the roof).[5] Or if everyone on Earth (about 5 billion people) counted at the rate of one atom every second, it would take them 4 million years to count all the atoms in a mole.[6] Or a personal favorite: if a million dollars had been given away each second since the beginning of Earth some 4.5 billion years ago, only about a quarter of a mole of dollars would have been given away in the history of Earth.[7]

So going in, Perrin knew he was dealing with very small targets and very big numbers, so direct methods just weren't going to work. But Perrin had the knack of standing back from a problem and considering it from a novel and unprejudiced perspective. Jean Perrin was, after all, a free thinker.[8] In his philosophy, he was agnostic. In his politics, he was socialistic. In his science, he was objective. When he went to weigh the atom he unapologetically and uninhibitedly turned to a phenomena that was far outside his field: the dancing pollen grains of Robert Brown.

Robert Brown (1773–1858) was a botanist and one of England's greatest. He served as librarian to the president of the Royal Society.[9] When offered a university position, he chose to stay at the library because it allowed him access to the glorious collections of the society.[10] He is known for his first observation of plant-cell nuclei and for the classification of several new plant species. But physical chemists should always think of him in conjunction with Brownian motion.

A careful and inquisitive investigator, Robert Brown subjected the pollen of his new species to microscopic examination. But he had difficulty with some very small pollen grains. When he placed them with a drop of water on the microscope slide the little pieces of pollen just wouldn't stay still. Naturally he found this a curious phenomena because he was aware that plant pollen should be inanimate, but the small particles didn't seem to know this for themselves.

What followed was a lot of effort by Robert Brown and several others to calm these bobbing corks. Brown himself tried treating them with alcohol and heat to kill any microscopic life that might be causing the animation. When this didn't stop the pollen from moving, he ground up small pieces of material, which he knew had to be dead, to see if they bobbed about, too. They did. Rock dust did. Soil dust did. Brown even combed the archives of the society for dust from a sphinx assembled thousands of years ago.[11] It bopped about, too.

Maxwell stated pragmatically that the movement must be an instrumental artifact or there must be an external cause, then let the matter drop. But others were not so easily assured and tried to discern the source of this strange movement. They found that shielding the particles from stray light and purposefully illuminating them with varying intensities and colors of light did not affect the movement. Allowing the system to rest long periods of time did not stop the movement, or even slow it down. Running the experiment during the day in the city revealed the same movement as running the experiment at night in the country. But what struck Perrin was that the movement was not regular, such as particles being carried by current, but that each grain bobbed independently, as if being bopped about by invisible fairies playing a minute game of keep-the-ball-aloft. The fairies, said Perrin, were atoms, and the balls were kept aloft by the perpetual movement assumed in the kinetic molecular theory.

There were other scientists, of course, who speculated that the motion of the pollen grains might be caused by the agitation of atoms, but Perrin was an heir of Lavoisier. He knew that a theory without numbers is not good enough. A profitable theory needs to make quantitatively verifiable predictions as well as qualitatively describe observations. He was a true physical chemist, and one of the best of his breed.

For all intents and purposes, the birth of Jean Perrin (1870; he died in 1942) corresponded to the formal birth of the discipline of physical chemistry.[12] Ostwald started his career a mere twenty years ahead of Perrin, and by the time Perrin came on board, physical chemistry was the hot new area of research. Jean Perrin was of unpretentious origins. His father was an army officer who died of battle wounds, so his mother raised Perrin and his two sisters. His academic talents, through scholarships, supported his education. Even with the scholarships, he had to take time from his studies for the military, which resulted in delaying his doctorate until he was twenty-seven. Once back in academics he was assigned the task of organizing a new course in physical chemistry at the Sorbonne and began writing a physical chemistry textbook. He married a woman with a university degree, which at that time was still a fairly unusual accomplishment, and they had two children, a girl and a boy.

Perrin's associates at the Sorbonne included Marie and Pierre Curie as well as other notables in the history of science.[13] In this enlightened atmosphere, however, the word *atom* was still barely spoken; the term *equivalent* was used instead. Perrin, however, was an open advocate of the atomic model and kinetic molecular theory. Perrin's first experiments on molecular motion were grounded in Maxwell's analysis of the pressure gradient of the atmosphere, which was based on the kinetic theory of gases.

In his analysis Maxwell showed that assuming a gas is composed of atoms in constant thermal agitation leads to the prediction of a pressure gradient in a column of gas: The pressure should be higher toward the bottom and less toward the top. This gradient derives from the tendency of the particles to move in all directions, including up, and the effect of gravity pulling them down. Maxwell also concluded that the exact pressure change could be calculated if one knew the temperature and the mass of the individual gas particles. However, working backward—that is, calculating the mass of the individual gas particles from the pressure difference—is not a practical experiment: To have a pressure difference significant enough to be measured reliably, say a 10 percent difference at room temperature, one would have to have a column of gas over 700 meters (about one-half mile) tall. Even if such a column were constructed, it would be another problem to try to keep the temperature constant over the entire length of the column. The column would have to be filled with a pure sample of gas, otherwise the mass measured would be the average mass over several different types of gas molecules. And, of course, one would want to check one's results by making the measurement several times at different temperatures and several different pressures—so this would be a truly ambitious project, especially considering that the very existence of atoms was still in doubt.

Perrin, however, saw a solution. Perrin suspended tiny pollen-like particles in water and measured their vertical distribution. Though tiny,

Fig. 17. Francis Henri Jean Siegfried Perrin, ca. 1948. AIP Emilio Segrè Visual Archives, Physics Today Collection.

these particles were many times more massive than air molecules, so the gradient in their distribution was apparent over a much shorter distance, just as the vertical distribution of sand in a wind storm is less than the vertical distribution of leaves. The particles he used were from a material called gamboge, a yellow, tropical-tree latex that was traditionally used for a water-color pigment. The yellow color helped with visibility and the low density of the material made low-weight particles that were still large enough to be seen. Using a microscope, Perrin measured the height distribution of the particles in a drop of water.

After accounting for the buoyancy of water, the mass per mole of gamboge was derived from the observed vertical distribution. Perrin measured the radius of the particles and from this could calculate the volume per particle. By measuring the density of the dry gamboge he arrived at the mass per unit volume. Putting these three pieces of information together—the mass per mole, the mass per unit volume, and the volume per particle—enabled him to calculate the number of particles in a mole, Avogadro's number.

Absent from the above description of Perrin's work, however, is a flavor for the excruciating exactness required for its execution, the elegance of the experimental design, and the heroic patience and persistence of Perrin. We are obliged to take a moment to consider these.

To begin with, Perrin not only had to make very small particles of gamboge, he had to make them of uniform size. Initially he produced these particles by rubbing the gamboge under water, like soap, then dissolving the foam in alcohol and centrifuging it to segregate particles by mass. He then dried them and examined them for uniformity. The amount of work this required is evidenced by the fact that it took several months to isolate a few tenths of a gram of suitable gamboge from the kilogram from which he started.

Then, employing a new advance in microscopy that illuminated the sample from the side rather than the bottom, he measured the number of particles in successive layers by changing the depth of focus by a millionth of a meter each increment, using a screw-type micrometer to move the lens. But the task remained daunting because the particles swam around so much it was difficult to count them. He was able to project the microscope image and take photographs of the larger particles so that the diminishing distribution with height could be clearly seen and evidenced in his publication. But to get the most accurate results he had to use the smallest possible particles, and these did not photograph well. Deciding that the problem was that the particles swam in and out of his field of vision before he could count them, he narrowed the field by covering the eyepiece with a piece of foil that had a hole he made with a dissecting needle. In this way he saw only a few at one time, so he could count them quickly. This was necessary because he took readings every fifteen seconds. For the first four papers he did all of the work himself, employing a graduate student only on the fifth paper.

After calculating Avogadro's number from the vertical distribution of gamboge particles and finding that his number agreed well with the estimates, he began to cast about to see if more information could be extracted. After so many hours of staring at the tiny particles, he became convinced he was observing particles being displaced as the result of collisions, not particles moving with a definite velocity. He was considering ways to relate his findings to the kinetic motions of atoms, when a colleague brought to his attention the work of Albert Einstein.

While Einstein is obviously best known for his theory of relativity, he also did considerable work on the kinetic molecular theory of gas, the hot topic for physicists of that day. Because we have the luxury of drawing on a great deal of knowledge of Einstein's life, let it suffice to say that he was still in the patent office when he considered the kinetic molecular theory and how it related to Brownian motion. Basically Einstein said that if a particle is in a sea of molecules in thermal agitation, then its net displacement in any one direction should change as the square root of time. Einstein's derivation was elegant and algebraic, but will be explained here by analogy with dirty clothes suffering displacement on laundry day.

On laundry day, in a typical household, clothing is sorted into stacks for separate batches in the most convenient wide-open space, such as the kitchen. Imagine for a moment piles of dark colors, light colors, whites, blue jeans, towels, sheets, and socks. Now imagine children running randomly through the kitchen, kicking the piles. As can be imagined, the piles will start to break down and the individual pieces of laundry will start to migrate. Now for a moment consider one red sock on the pile of socks and how it might move toward some arbitrary direction, for instance the living room. The first kick is guaranteed to displace it from its original position and the displacement may be partially or wholly in the chosen direction. But though the next kick might move it farther toward the living room, it might also move the sock back toward its original location. Some subsequent kicks will move it farther, but some will move it back, too. With these random kicks, the sock slowly moves closer to the living room, but this is much different behavior than if the sock had a definite velocity. If it had been picked up and thrown into the living room it would have arrived there in a much shorter period of time. Perrin carefully observed the way his particles moved and noted their displacement with time. He found the average displacement varied as the square root of time, which is how the particles would move if they were being randomly kicked by atoms. He then used Einstein's equation based on displacement to calculate Avogadro's number again. He found that the second number agreed well with the first. Therefore the assumption of the atomic structure of matter not only predicted one measurable, but two, and Perrin had measured them both. If these observations weren't the result of atoms, it would be hard to explain where they did come from.

When Einstein received a copy of Perrin's paper, he commented, "I had thought it impossible to investigate Brownian movement so precisely. . . . In my eyes it was only a nice little game."[14] Though Germany and France would soon be pitted against each other (Perrin would serve France in the engineer corps during World War I), these two men corresponded amicably and eventually met to shake hands.

Even Ostwald, who had for so long distrusted the atomic hypothesis as unproved and unprovable, stated, "I have satisfied myself that we arrived . . . at the possession of experimental proof for the discrete or particulate nature of matter,"[15] citing the work of Jean Perrin. An acclamation, but a touch reserved. Other commentators were not so reserved and some were downright critical. Le Châtelier remained skeptical and Mach, who had issued the challenge, "Have you ever seen one?" seemed impressed, but still, in his last published statement, called atomism "hypothetical-fictive physics."[16]

Why didn't the entire community of chemists and physicists immediately rise to applaud the evidence that provided for such a useful and explanatory model? Because there were still pieces of the puzzle that did not fit, especially those confounding results obtained when light interacted with matter—the purview of the maturing field of spectroscopy. But Perrin had shown that if you cannot see the cause, you might see the effect and from this deduce the cause. Though he would have to flee his beloved Sorbonne ahead of the German invaders in World War II and would die in the United States a few years later, Perrin would live to see this dictum applied to spectroscopy and from its application the final vindication of his atoms. Spectroscopy, as Perrin understood, is another way to see the unseeable. At the turn of the century, spectroscopy would shed light on a whole new world.

Part IV

Physics and Chemistry Come to Light

INTRODUCTION

> *The wish to capture [photographic images] . . . is blasphemy. . . . Is it possible that God should have abandoned His eternal principles, and allowed a Frenchman . . . to give to the world an invention of the Devil?*
> *Leipzig City Advertiser*, circa 1830

While it is true that the focus was on materials during the Industrial Revolution, there was, at the same time, another group struggling with the nature of light. These two groups used different models (particles versus waves), different mathematics (algebra versus calculus), and different instruments (microscopes versus telescopes), but they found common ground in their curiosity concerning energy.

195

Gradually, as the range of light extended into the invisible and the electromagnetic nature of light became understood, the two foci merged, and quantum mechanics resulted.

A technological advance would come from the interaction of those studying light with those studying matter: photography. This technique, it may be recalled, was used by Perrin to illustrate the distribution of his gamboge particles. It would be used by astronomers, biologists, and physicians as well as chemists and physicists to augment their science. The development of photography required a blend of physics and chemistry.

On the physics side was the camera obscura, a dark box or room that had a small hole on one side that allowed light to enter. Images illuminated by light outside the box appear on the wall in the box, opposite the hole, with the image inverted. European medieval artists used it as an aid to drawing. Our friend the medieval magician, Giambattista della Porta, was fascinated by it. He added a lens to the hole and wrote extensively of the effect.[1]

On the chemistry side, by the 1600s many people had noticed that certain chemicals bleached or darkened over time; they tended to attribute the effect to heat or exposure to air, not light. Angelo Sala in the mid-1600s noticed that silver nitrate blackened in the sun.[2] Elizabeth Fulhame, in the mid-1700s, obtained patterns by soaking cloth in gold salts and exposing them to the sun. She wrote of her results in a book intended to support Lavoisier's theories concerning the oxidation of metals. Undoubtedly Lavoisier would have appreciated her support, but he died on the guillotine six months before the book was published.[3] In the 1800s, Louis Daguerre and J. Nicéphore Niepce found a method for fixing images on a silver-coated copper plate.[4] In 1908 Gabriel Lippmann was awarded the Nobel Prize in physics for his method based on the interference phenomenon for reproducing colors photographically.[5]

As late as the 1800s fallout from the Protestant Reformation was still being felt. Germany, the main battleground of the Thirty Years War of the mid-1600s, had suffered dismemberment into small principalities that an emperor loosely held together. The leadership of Otto von Bismarck eventually brought Germany under a strong central government and made Germany a European power. However, Germany found itself behind in the Industrial Revolution. The result was increased German interest and support for science and technology, but the race to catch up fostered a rivalry among Germany, England, and France. The divisiveness in politics led to divisiveness in science, which led to divisiveness in politics. In the arena of science, competing theories, such as the English and German theories of light, would emerge. In the political arena, the rivalry would culminate in the lightning bolt of the world wars. But before that time, for a while at least, Europe would bask in the light.

Chapter 17

Spectroscopy

The sun . . . passeth through pollutions and itself remains as pure as before.

Francis Bacon, circa 1600

While still debating the question of atoms, the question of light for many scientists was settled. Newton had used different models for light, particle or wave, depending on his needs of the moment,[1] but he was also quite comfortable with transmutation and action at a distance. Most scientists decided that Newton, in the main, had declared for the particle model and therefore so should they. Later, in the early 1800s, experiments were done that seemed to define light as a wave. These were again so convincing that the pendulum swung the other way. Such was the situation when there arose a breed of scientist that solved the problem by not worrying about it. These *spectroscopists*—people who study the interaction of light with matter—had as much in common with the botanists as the physicists.

They were more interested in collecting and cataloging the beautiful colors and patterns from the sun than in tethering their light with theories. These practitioners of the art knew that eventually the theorists would have to decide where the colors came from. But, for the moment, they saw as their mission the gathering of information for the theorists to digest. They were collecting rainbows from the sun.

The word *spectrum* is commonly used to denote a range; for instance, one might speak of the broad spectrum of human behavior or the narrow spectrum of socially acceptable behavior. Today we know that the optical spectrum is a range of energy: Some light is high energy and some is low energy. The energy of light from a rainbow or a sunset is registered as color by our eyes, but there is also light that has energy outside the visible range. This light may not be seen, but may be felt, like heat from a stove burner, or can cause change, such as sunburn from ultraviolet rays. The discovery that light comes in a range of energies rather than just one was made by Newton. He broke up light into its various colors with a prism, which was not new, but then had the inspiration to use another prism to recombine his rainbow into white light, thus demonstrating the action of the prism was to separate and combine light, not to change it. It took another hundred years to decide that there was more to the spectrum than meets the eye: light that could not be seen.

The first indication of light outside the visible range was found by someone whom we've encountered earlier, in reference to the work of his son, John Herschel. We applauded John Herschel for introducing statistical techniques to England and we mentioned that he had grown up in a stimulating intellectual household where he and his aunt/tutor Caroline assisted his astronomer father, William Herschel (1792–1871).[2] William Herschel, who would become Sir William Herschel, came to England to escape the French occupation of Hanover and originally supported himself as a music teacher and organist. One of the father's research programs resulted in the discovery of the planet Uranus and another resulted in the detection of light outside the visible range. In 1800 he measured the warming power of different colors of prism-separated light by placing a thermometer in a beam of each color light. He put a thermometer just outside the red end of the spectrum as a control and found it still reg-istered warmth. The invisible light that was the source of this warm-ing came to be known as infrared light. Infrared light is the light given off by animals and that can be detected with infrared goggles.[3]

Johann Ritter (1776–1810) was likewise of humble beginnings. He was born the son of a Protestant pastor in what it now part of Poland. Originally an apothecary's apprentice, he studied chemistry texts and did experiments on his own, even to the detriment of his apothecary work. He used his small inheritance to attend a university and study electrochemistry. When he gained a measure of success in this field he married the young woman with whom he had been living and by whom he had a child and they subsequently had more chil-dren. But fame did not bring happiness, as he was quarrelsome and tended to relocate, probably because of the quarrels. When he heard of Herschel's discovery of energy at the red end of the spectrum, he imagined an analogy with the two poles of a magnet and looked for energy at the other end of the spectrum. In 1801 he found it. Silver chloride, a white crystalline solid that blackens in the presence of light, blackens even more rapidly in the region of the spectrum beyond violet where there is no color discernible to the human eye. Buoyed by his success he began to investigate other curious phe-nomena, including occult traditions such as water divining. He came to believe that magnets could influence the operation of the human body and invented the word "siderism" to describe the study of such effects. However these investigations fell so much outside the estab-lished range of scientific inquiry that despite his discovery of the light beyond the violet end of the spectrum—ultraviolet light—he was never called to a university position.[4]

More progress was made by Joseph von Fraunhofer (1787–1826), the eleventh and last child of a family of German artisans in the glass and optical trade, who dreamed of being a master lens maker.[5] Unfor-tunately his parents died when he was twelve and he was apprenticed by his guardian to a mirror maker who wanted him to make mirrors and not further his education. Then, in a twist of fate, his master's workshop collapsed and Fraunhofer was pinned in the wreckage for several hours before he could be freed. A local German prince heard of his misfortune and in a fairy-tale–like gesture gave the boy a sum of money sufficient to allow him to buy books on optics, a glass-working machine, and eventual freedom from his apprenticeship.[6] In his quest to understand the workings of lenses and especially chromatic aberrations (unwanted colored fringes seen when using inexpensive binoculars), Fraunhofer found dark lines interspersed among the colors of the solar spectrum. Others had seen these lines, too, but Fraunhofer, knowing optics so well, was able to find ways to expand the spectrum and make the lines even sharper. Eventually he observed and cataloged some 500 lines in solar light. His published findings were so well received, the dark lines became known as Fraunhofer lines. But they were regarded for the moment as a curiosity rather than the useful tool they would eventually become.

Around 1825 an English gentleman scholar, William Henry Fox Talbot (1800–1877), joined by John Herschel (1792–1871), began investigating the different colors that metal salts lend to flames (just as metal salts are used to add color to fireworks).[7] Particularly bright were the sodium, potassium, and lithium flames. They set a prism in front of these flames and found that each of the different metal salts produced a distinct, characteristic spectrum that was not a continuous rainbow, but a set of bright lines at very specific locations with respect to each other.[8]

In 1832 this observation was elaborated on by David Brewster (1781–1868), a self-educated Scottish physicist.[9] Though Brewster held no formal degrees, he had served as secretary to a gentleman scholar and had assisted in the assembly of microscopes, telescopes, and sundials. He invented the kaleidoscope and made substantial contributions to the field of optics,[10] including investigations of refractive index, the property of light used by H. G. Wells in his 1897 *The Invisible Man*. Refraction is the "bending" of light that takes place when the light travels from one medium to another, and the refractive index is a measure of the light-bending properties of a material. The bending of light is why you have to reach in front of the apparent position of the pebble when you pick a pebble from a stream. The refracted image makes the pebble seem like it's somewhere it isn't. In *The Invisible Man*, Wells has a chemist discover a material that will change the refractive index of skin so that instead of reflecting light it allows light to pass through. That this chemical might at least be philosophically possible can be demonstrated by wetting a paper towel and laying it on your hand. Dry, you cannot see through the paper towel. Wet—that is, after adding the chemical water—you can see your hand. Of course *The Invisible Man* strains credulity by suggesting that the chemical could affect the refractive index of all organs equally, but that's what makes it science fiction and not science.

Brewster's place in the history of spectroscopy results from the experiment he performed in which he passed white light through various gases. He saw that lines of light were subtracted from the white light as it passed through the gas and this "dark spectrum" was characteristic of the gas: Each different gas had its own pattern of lines. Several scientists then put forth the idea that the patterns of lines could be used to identify elements, but suggesting and proving, however, are two different things. It required the studied eye and patience of a German theoretical physicist and a German experimental chemist to make the exacting measurements that turned the idea into a viable tool. The physicist was Gustav Kirchhoff (1824–1887) and the chemist was Robert Bunsen (1811–1899).[11]

Kirchhoff, described as a rather private person, had suffered an injury early in life and had to use crutches or a wheelchair to perform research or teach. A member of a family of the German intelligentsia with loyalty to the Prussian state, he was nearly guaranteed a successful academic career when he completed the prescribed course of study and then married the daughter of an influential academic. He met Bunsen when Bunsen came to Breslau, the university at which Kirchhoff taught. Although Bunsen stayed only two years in Breslau, it became apparent to the two men that they could form a successful collaboration. Bunsen was a very practical man and so devoted to his teaching and his research that he never married.[12] He apparently was engaged at one time, but his intended suffered from such neglect that the engagement was broken off.[13] In his teaching he focused on fact over theory and did not even present Avogadro's hypothesis nor the periodic table, though it was developed by his own students, Mendeleev

Fig. 18. Gustav Kirchhoff. Edgar Fahs Smith Collection, University of Pennsylvania Library.

and Meyer. He was a hands-on researcher who made his own apparatus, including glassware, and unfortunately lost an eye to experimentation with an explosive substance.[14] He was interested in the use of spectroscopy as an analytical technique, not particularly in its theory. Yet something of the theoretician Kirchhoff appealed to the technician and something in the technician appealed to the theoretician. In 1854 Bunsen arranged for Kirchhoff to be called to Heidelberg.

Bunsen was trying to distinguish different materials from the colors of their flames when Kirchhoff suggested that Bunsen pass the light from the flame through a prism and look at differences in the spectrum of the flame instead. They then engineered a method for studying the spectrum carefully and reproducibly by using a permanently mounted prism, scope, and a new type of device for producing a flame known today as the Bunsen burner. Anyone who has been in a chemistry laboratory for any amount of time has used this device. It allows natural gas fuel to mix with air from an intake that can be regulated. When properly adjusted the result is a clean, hot, steady flame. They purified their samples carefully and viewed the spectrum of the flame at an angle. With this instrument, they observed that indeed each element had its own characteristic set of fireworks-like lines. For instance, sodium produces characteristic yellow lines; mercury produces two violet, a blue, two green, several yellow, two orange, and a red line, the relative positions of which are as identifying as fingerprints. They devised a scale that allowed the positions of the lines to be catalogued carefully. By early 1861 the team had used the spectroscope to identify two new elements: cesium (from the Latin for sky blue) and rubidium (from the Latin for deep red).

Kirchhoff got to playing around with the apparatus (which had earned the name *spectroscope* by now) and viewed sunshine through a sodium flame. What he found surprised him. When the sunlight was intense enough, the characteristic sodium light disappeared and was replaced by a dark line. He concluded, after some thought, that elements must absorb the same light that they admit. When the debate began as to the source of atomic interactions with light, he was able to add a pivotal piece of information based on his observations. For the moment, however, he used it to suggest a very practical result: The spectrum of the sun contained a dark line in exactly the same position as the one produced by burning sodium, so the sun must contain sodium in its atmosphere. However the characteristic lines for lithium were missing, so no lithium was in the atmosphere of the sun.

One can only imagine the excitement this discovery engendered. Here now was a method of discerning the chemical composition of the stars. Not only were the heavens no longer Plato's perfect orbs, they were chemical and knowable through these instruments. The romance of the notion did not escape one man in particular: Joseph Norman Lockyer (1836–1920). In 1868 he examined the solar spectrum and decided that he could see a new substance that had not yet been identified on Earth. He named it helium for the Greek *Helios*, the sun god, and helium remains the only element that was identified extraterrestrially before being found on Earth.[15]

Lockyer's pronouncement, however, did not meet with rounds of applause because the scientific community had become dulled to pronouncements by Lockyer and had developed quiet a shield of skepticism toward them. For instance, Lockyer had decided that atoms dissociate in flames to hydrogen and other elements, but other chemists had been able to show that his samples were in fact contaminated with hydrogen. He also believed planets are condensations of swarms of meteorites and there was a correlation between sunspots and Earth's weather. However in 1895 Sir William Ramsay found and identified helium from a terrestrial source and vindicated this one discovery of Lockyer. Lockyer had another legacy worthy of note. In 1869 he founded the journal *Nature*, which he edited for the first fifty years of its existence. The juxtaposition of the date of the founding of the journal (1869) with his pronouncement for helium (1868) makes one wonder if he did not perhaps create this journal as venue for his own ideas. At any rate, the journal has now become perhaps the most prestigious scientific journal in the world presenting a broad, rather than specific, coverage of science. In all, Lockyer was probably best described by his biographer Herbert Dingle as an "adventurous rather than critical scientist."[16]

But then adventure was the watchword of the day. In the 1890s, William Henri Julius (1860–1925), a colleague of Einstein, found that organic compounds absorbed infrared light and that the portion of the infrared spectrum they absorbed was characteristic of the compound. He used invisible light to probe invisible molecules.[17]

But while the spectroscopists happily collected, photographed, and cataloged, a storm was brewing. Indications that there was more to the spectra that needed to be understood came when Lecoq de Boisbaudran (1838–1912) discovered the elements gallium, dysprosium, and samarium in his home-built laboratory by noticing that there were regularities in the spectra of certain elements—patterns—that were correlated with their atomic weights.[18] Moreover, in 1885 Johann Jacob Balmer (1825–1898), at the age of sixty, found an arithmetic formula that predicted the spacing of the four lines of a published hydrogen spectrum. This Swiss secondary school teacher and father of six had earned a doctorate in mathematics, but had never been called to a university position. He told a friend who was a university professor about his formula and this friend told him that there was a fifth line that had been identified. Balmer's formula matched the fifth. And then more.[19] One can only imagine the feelings of a sixty-year-old schoolteacher on finding that he had discovered a fundamental new mathematical pattern in nature. True, his finding was to drive the spectroscopists from their garden out into the rigors of theoretical interpretation. But all in all, it may be believed that they had enjoyed their moment in the sun.

Chapter 18

Electromagnetism

I have not had a moment's peace or happiness in respect to electromagnetic theory. . . . I have been liable to fits of ether dipsomania, kept away at intervals only by rigorous abstention from thought on the subject.
Lord Kelvin, circa 1896

Before the interaction of light with matter could be understood—the interaction that gave rise to Johann Balmer's beautiful lines—light had to be understood, and matter had to be understood. We will now walk with the physicists and chemists who determined light was a wave and then follow them as they discover exactly what was waving. In this journey we will have many heroes, but two in particular: Michael Faraday and James Clerk Maxwell. Faraday, the chemist, made the essential observations and had the original concepts that allowed Maxwell, the physicist, to synthesize a mathematical model of light. In regard to Kelvin's quote above, it was Faraday and Maxwell that caused Kelvin's condition.

Early evidence for the wave nature of light came from Thomas Young (1773–1829).[1] Though raised an English Quaker, Young had a taste for music, dancing, and the theater. Well married and supported by a wealthy uncle, Thomas Young was a classic example of a gentleman scholar of the age. But while the phrase "gentleman scholar" may conjure up an image of indolence, in Young's case, at any rate, this was far from the truth. In addition to studying medicine and physics, he made a study of ancient Egypt and contributed to the

207

deciphering of the Rosetta Stone. He also applied his talents to the study of light and the search for evidence to convince others of what he believed, that light was a wave, and he succeeded. Young devised a demonstration of light interference that convincingly argued for the wave model of light.

Interference is what happens when two waves meet. When two wave crests meet, they add to form a larger crest. When two wave troughs meet, they form a deeper trough. When a wave crest meets a wave trough, there is a calm. Wave interference can be modeled by two combs. If the combs are overlaid so that the teeth align, dark and light spaces can be seen between the teeth, which represents constructive interference. If the combs are offset so that the teeth of one overlay the gaps in the other, this models destructive interference: There are still two combs, but no wave.

Though considerably more difficult to execute in Young's day, the phenomenon of light interference can be demonstrated by anyone today with access to a laser pointer and a metal ruler with a raised scale, that is, one with bumps rather than painted-on lines. In a darkened room (the darker the better) the ruler is laid atop a pile of books or a box and the laser pointer is mounted on a flexible-necked study lamp so that the angle with which the laser light hits the ruler can be easily adjusted. The laser is turned on and the lamp position changed until a pattern of light and dark lines is seen on the wall opposite the laser. A low angle with respect to the ruler works best. The pattern results because light waves bouncing off the ruler crest together at some positions and give rise to a calm at others.

A laser pointer is necessary for the demonstration because it provides monochromatic light, or light of a very narrow band of color. Young, obviously, did not have a laser, but he did have access to fairly monochromatic light. He split white light into colors by a process called diffraction. Wave diffraction can also be demonstrated in a pool, a bathtub, or even the kitchen sink. If something with a grid of uniform slots, such as a slotted spatula, is put in calm water and a wave is generated on one side, it can be seen that as the wave passes through the slots to the other side, the wave bends so that the result is a circular wave on the opposite side of the slots. With light, the amount of bending will depend on the color of light, so when light diffracts around an object it will separate into colors as though going through a prism. Young used diffraction to separate his light by putting a small card of paper in front of a pinhole source of sunlight. The shadow of the card had fringes of color, as he expected, but in addition he noticed that the shadow itself was cut by fringes. Young knew that just the observance of diffraction was not enough to prove the wave nature of light. True, it could be explained by waves, but it could also be explained by particles. If light were formed from particles, like little bullets, then they should either go straight past the card or bounce back if they hit it, but at the edges they might ricochet enough to have their trajectories bend at strange angles. However Young took his experiment one step further. He held up a barrier so that the light from one side of the card was cut off. When he did this, the fringes in the shadow went away, which meant that it required light coming from both sides of the card to make the fringe in the center. This was not the behavior of bullets. If the ricocheting of the bullets were causing the pattern of fringes in the center, then the fringes caused by bullets ricocheting from one side should not change when the other side is covered. In other words, something happening in Tombstone should not change the way bullets ricochet three miles away at the OK Corral.

Even though Young's demonstration was powerful, there was still a need for a mathematical model with predictive power. In 1818 Augustin Fresnel (1788–1827), entered a competition sponsored by the Academy of Science of Paris, the theme of which was light.[2] He was a rather tragic figure who suffered in the political turmoil of Napoleon and the Restoration, who would die of tuberculosis at the age of thirty-nine. After many months of struggling with the very difficult equations that described his waves (made possible, but not simple, by Newton's calculus), Fresnel had a model to present. Siméon Denis Poisson (1781–1840) was a judge in the competition who was actually a proponent of the particle theory of light.[3] He realized that Fresnel's mathematics predicted that there should be a bright spot behind a disk held up in a beam of light because the light would bend around it. (The realization of this prediction is often credited to Fresnel himself, but it was Poisson who saw it.) Poisson challenged Fresnel to do the experiment, he did, and he saw the light. Although Fresnel was dismissed by Young as an opportunist cashing in on his (Young's) genius, it is now well accepted that Fresnel was working independently, and although he acknowledged Young's priority in the demonstration of interference, Fresnel's mathematical model was original.

But while this revolutionary work was proceeding, an annoying question kept arising: waves of what? If light was a wave, what was waving? The answer, when it arrived, was from a seemingly unrelated area of endeavor: electricity.

By 1800 Luigi Galvani had used electricity to make frogs' legs twitch, and Alessandro Volta had assembled his voltaic pile from acid and metals. Because the voltaic battery was such a convenient source of electricity and straightforward to assemble, soon every physicist and chemist was zapping everything that wasn't faster than they were. But electricity had always held a fascination for scientists and had been studied since the time of the ancient Greeks, who were aware of magnetism as well. (The Western name for both of these phenomena have their origins in the Greek language.) The Greeks and moderns were also aware that there must be some relationship between the two because they knew that lightning strikes sometimes cause metals to become magnetic. So Hans Christian Oersted (1777–1851), Danish natural philosopher and teacher, looked for this relationship between electricity and magnetism.[4] He knew one must exist, but he was stuck in a rut of preconception: He kept placing his current-carrying wire at right angles to the magnetic needle that he was trying to move and continued to be disappointed. Then one day, in an in-class demonstration, in front of an assembly of students, he found his battery still had some charge and he still had some time in his lecture. So he decided to put the wire parallel to the magnetic needle. The needle swiveled under the influence of invisible force and turned at right angles to the current. He published his finding in 1820, at the age of forty-two, and for the next thirty-two years played with his wonderful device. The phenomenon that Oersted discovered is one that is still routinely demonstrated by wrapping a wire around a nail and then running current from a low-voltage battery through the wire. The electromagnet assembled this way will pick up other metal objects and drop them when the current is cut off. Oersted had found that electric current induces a magnetic effect. The complementary phenomenon, that is, an electric current created by a moving magnet, was demonstrated in 1821 by Michael Faraday.

Michael Faraday (1791–1867) was a sort of antithesis to the gentleman scholar.[5] He was the son of a blacksmith who started his life as a common laborer, apprenticed to a bookbinder. His often repeated but still noteworthy story includes reading Mrs. Marcet's *Conversations on Chemistry* while working as a bookbinder and aspiring to a career in chemistry after hearing the notable chemist Sir Humphry Davy lecture.[6] Faraday wrote to Davy requesting any possible position, but received no reply. Not to be dissuaded, he sent Davy a bound copy of his own notes taken in Davy's lecture, and to this Davy replied. The next time Davy needed an assistant, he sent for Faraday.

Because Davy was a chemist, Faraday did most of his work in chemistry. Through his association with Davy, though, he became familiar with electricity, too. Davy had used electrochemistry to isolate the elements sodium and potassium from materials that Lavoisier had believed to be inert, primary elements.

Faraday soon showed himself to be a talented experimentalist in his own right and began making discoveries on his own. Unfortunately this led to some uneasy feelings between Davy and Faraday. At one point in his career Faraday had acted as valet to Davy, and the dissolution of the distinction between classes was not easy to achieve in nineteenth-century England. A case in point was Faraday's successful demonstration of the induction of electric current by a magnet. Davy, working with another notable chemist, had tried to demonstrate the connection between electricity and magnetism that they knew intuitively must exist, but they were unsuccessful. While they were on holiday, Faraday got into the lab and made a device that worked. In 1821 Faraday succeeded in making a current-carrying wire rotate around a magnet and visa versa: electromagnetic induction. Davy was not pleased.

But the rest of the world became excited. Electromagnetic induction is the effect that is responsible for virtually all the electric power generation that doesn't derive from batteries. So even though Davy was not pleased, Queen Victoria presented Faraday with a house to live in when he retired.[7]

Like a fine race horse tethered to a plow, Faraday had a spark of genius that belied his humble circumstances. Had he had the benefit of an early education in mathematics and more leisure to cultivate his talents, he may have accomplished even more. Faraday was more than just experimentally clever. He also thought about the effects he witnessed and their nature and origin. The depth of his thoughts rivaled those of Newton.

As we may recall, when Newton proposed his equations for gravity, there were those who were uneasy about the idea of action at a distance. How could the moon know it was supposed to be attracted to the Earth? If the Earth were to disintegrate spontaneously, would the moon instantaneously drift free? According to Newton it would. If there were no more Earth would its tug of gravity also instantaneously cease? According to Newton it would. And now here in electromagnetism this troubling action at a distance seemed to come up again. But these ideas ran against the grain of most experienced scientists. They felt there had to be something causing the communication between Earth and moon. The ancient Greeks had postulated an additional element, *ether*, to fill the space between Earth and the moon, and now others postulated invisible filaments or particles bouncing back and forth. But neither the ether nor the particles could be caught or weighed or measured. For electromagnetism, however, Faraday found something that could. As everyone who has played with magnets knows (and those who haven't played with magnets are hereby invited to do so), iron filings will form a very distinctive pattern around the periphery of a magnet. Here then was something visual and measurable. Faraday described it as lines or tubes of force and it has become known as a *field*. It would take some time to extend this notion to gravity, but for terrestrial electromagnetism, at least, there was now a way around action at a distance.

The notion of a field is useful in many areas of physical science. A

Fig. 19. Faraday's lab in the Royal Institution. Edgar Fahs Smith Collection, University of Pennsylvania Library.

room with a central fireplace (or stove) has a temperature field. You can measure the field by measuring the temperature at different points in the room. Our bodies sense temperature fields but they do not sense normal electric fields, though they are there. A shark, on the other hand, can sense variations in the electric field through special sense organs on its skin.[8] Similar to a temperature field set up by a fireplace, an electric field radiates out in all directions like a balled-up porcupine and falls off with distance from the source. Faraday identified both electric fields and magnetic fields and thought of electromagnetic induction in terms of fields. The value of the concept of field, in addition to being visual and measurable, was that when the field is set up by an electric charge or a magnet, it exists whether another charge is in the locality or not. (Remove the stove and you still have a temperature field.) Thus it is not necessary to invoke action at a distance. The field is there and the force of the field is doing the communicating.

Faraday also had some fundamental conceptions about matter: He knew from his own experiments that there was a magnetic field associated with the electric field; he knew electricity could decompose matter; and he knew matter could polarize light, so he felt in his bones there must be a connection, and he looked for it. In 1845 he found it: He observed that a magnetic field could rotate the plane of polarized light.

Among the Holy Grails of science, along with solving the human genome (done), finding an unlimited power source (pending), and finding a cure for all cancers (pending), is a goal called the Grand Unified Theory (GUT), which is the hope that a theory might be found that will unite all of the fundamental forces—gravity, electromagnetism, strong internuclear forces, and weak internuclear forces. The first inkling that such a goal was even possible was Faraday's unification of the fields of electromagnetism and light.

Though Faraday's fields were difficult for many to come to grips with, not all were so shaken. For one, there was the unflappable Maxwell, who had calmly considered the rings of Saturn and dared to believe the inverse viscosity effects of gas. When he read of Faraday's lines of force, he saw the light. Beginning with his 1856 paper "On Faraday's Line of Force," Maxwell began to publish a mathematical synthesis of the ideas and observations of Faraday.[9]

Maxwell saw the world thus: A moving charge sets up a changing magnetic field and a changing magnetic field induces a changing electric field to its front. The electric field creates a magnetic field, which creates an electric field, and this ripple effect, which has been described as leapfrogging,[10] will perpetuate ad infinitum if there were nothing to absorb its energy. Energy from the fields is exchanged, not depleted. Then Maxwell took the idea further. In 1862 Maxwell used known data to calculate the time it would take for a magnetic field to induce a matching electric field and propagate itself forward. He found that this speed matched the known speed of light. Maxwell had found the nature of light: oscillating electric and magnetic fields, each field producing another field ahead of it, at a speed of light.

Defining light as an oscillating electromagnetic field also helps us come to better terms with light interference. The cancellation of two opposing magnetic fields can easily be demonstrated with two bar magnets. Each magnet individually may be a strong magnet, but when they are joined, north pole to south, the resulting combination has greatly reduced magnetic power (although there will probably not be complete cancellation due to imperfections in the magnets). The same may be imagined for electric fields.

Maxwell, dead in 1879 at the age of forty-eight, did not live to see the experimental validation of his theory, which had to wait some twenty years until a spark source of a high enough frequency was developed. In 1888 Heinrich Rudolf Hertz (1857–1894) used such a spark to produce a changing electric field, which, according to Maxwell, produced a magnetic field, which produced an electric field, and so on, across the room.[11] Hertz's problem became the detection of the wave once it was generated. His inspiration was to use the same phenomena that generated his waves to detect them: He set up a gap between two metals similar to the gap in the spark generator and waited to see if the incoming electromagnetic wave would stimulate a spark in the receiver. It did, but one can only wonder at the patience and faith in ultimate success that Hertz displayed at this point.[12] To generate his spark, he used essentially the same setup that is used to generate a spark in a spark plug of a car. His detector was basically a radio receiver. But Hertz knew little about automobiles and nothing about radios. Karl Benz was busy building the first successful internal-combustion engine at almost exactly the same time as Hertz's experiments and Hertz was, in essence, inventing radio.

But have we moved far afield from physical chemistry with all this discussion of light and electromagnetic waves? Not at all. The essence of physical chemistry is discovering the fundamental character of materials and their interactions, and these are electromagnetic in nature. The most powerful tools in the physical chemist's store are the interactions of light with matter—spectroscopy—as well as electrical and magnetic probes. To appreciate what makes these tools so important, it is necessary to understand the fundamental structure of matter, the discovery of which we follow next.

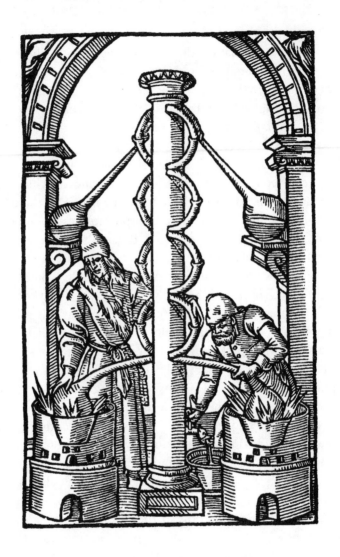

Chapter 19

Atomic Structure

*One day Rutherford, obviously in the best of spirits, came into my room
and told me that he now knew what the atom looked like.*

Hans Geiger, circa 1910

After the Nobel Prize was inaugurated in 1901, many of the
prizes in physics and chemistry were awarded for research on the
structure of atoms. One of the first of this crop of Nobel Prize win-
ners was J. J. Thomson. He achieved many of his results playing with
the century's new toy: the cathode ray tube. Though these cathode
ray tubes would eventually be used in televisions and computer monitors, at the time, they were little more than a novelty to be studied.
But from the study of these lovely glowing tubes would come information on the structure of the atom.

An atom is built from three basic components that determine
chemistry: electrons, protons, and neutrons. These components
themselves are made from their own pieces, but these bits of bits do
not change during the course of chemical reactions, and can be considered chemically moot, though they do exert a fascination of their
own. In the most simplified picture of the atom, the electron can be
seen to travel in a sort of orbit about the nucleus, which is made up
of the protons and neutrons, clumped together in the center.

Electrons are the negatively charged particles that flow through
current-carrying wires and are supplied by batteries or appliance
chargers. For instance, the information on a charger of a cellular

217

phone may say it supplies a current of about 250 milliamps. This current corresponds to about 10 million billion electrons per second flowing through the wires. We don't see great lumps bulging out from these wires, however, because electrons are very fast and very tiny. In fact, the electron has a rest mass of almost one millionth of a billionth of a billionth of a kilogram. Why "rest mass"? Because when an electron isn't holding still—which it never really does—it is moving at speeds comparable to the speed of light, and because of relativity effects, this means the mass varies with speed.

The neutron is an electrically neutral particle found in the nucleus, or center, of the atom. It doesn't contribute to the charge, but it does contribute to the mass. The proton is a positively charged particle also found in the nucleus of the atom. When the number of electrons and protons in the atom are balanced, the atom is electrically neutral. The proton is more massive than the electron, by a factor of about 2,000, as is the neutron, so the mass of the atom is determined by the protons and neutrons, though—like fleas on an elephant—the electrons are still very important. The mass of the elephant is determined by the elephant, not the fleas, and the elephant weighs the same with or without fleas, but the elephant is certainly aware of the fleas. The electrons determine the chemistry of the atom just as the fleas might determine the behavior of the elephant—they just don't contribute to the mass. So how did Michael Faraday manage to get his hands on these slippery fellows? The answer, of course, is that he didn't. What Faraday found were the tracks; it would be others who would bag the beast.

In 1832 Michael Faraday was investigating *electrochemistry*, that is, using the action of electricity to induce chemical change or using chemical change to produce electricity. Electroplating, the process by which jewelry is gold- or silver-plated, is an example of using electricity to induce chemical change. A battery is an example of using chemicals to produce electricity. Both of these processes come under the heading "electrochemistry." Specifically, Faraday looked at the breakdown of salts by the action of electricity, the research that earned his mentor, Sir Humphry Davy, his initial fame. Davy, by wiring a battery to a pool of molten sodium oxide, managed to obtain pure sodium metal. He did the same for potassium. But after his initial successes Davy seems to have been kept out of the lab by his fame and its social demands. Faraday, the humble laboratory assistant who didn't have to worry about fame, was free to indulge his curiosity in the lab. Through patience, he found that it required a definite amount of electricity to decompose a definite amount of material, a relationship now known as Faraday's law. For example, hooking up the battery to his apparatus for a few minutes resulted in a certain amount of material and hooking up the battery for twice as long resulted in twice as much material. But this was not the result he had expected.

Chemists were aware that certain materials seemed to display greater or lesser degrees of *affinity*, or attraction, toward each other. For instance, calcium carbonate, or chalk, does not dissolve in pure water; it requires an acid solution to dissolve it. Therefore calcium and carbonate were thought to have a good deal of affinity for each other. On the other hand, calcium chloride is quite soluble in water, so calcium and chloride were thought to have less mutual affinity. Initially electricity was thought of as a force that disrupted the affinity between elements. But if electricity were somehow exerting a force that broke the affinity, then it should require more electricity to break up compounds with more affinity. But it didn't. When producing sodium metal, it required the same amount of electricity to produce the same amount of sodium, regardless of the compound it came from. Faraday was not a believer in atoms, so he expressed his findings in terms of equivalents. He found that the sodium-forming reaction required the same number of equivalents of electricity per gram of sodium, regardless of the source of sodium.

Other hints arose suggesting there might be something material, rather than a force, that caused the changes. It had been established by then that the strength of an electric field varies as the inverse square of the distance, as does the force of gravity. In other words, the farther apart a positive and negative charge, the less the attraction between them. This can be demonstrated by creating a bit of static electricity in adhesive tape. If a strip of adhesive tape is pulled briskly from the dispenser, it will have a static charge. If this piece is allowed to dangle from a table ledge and another piece of tape, also pulled briskly from the dispenser, is brought up to it slowly, there will be no interaction between them until they are very close; then the repulsion between them (two like charges) increases rapidly. But Faraday showed that distance did not matter when decomposing salts with electricity.[1] In addition, after the initial reaction, the material formed separated from the wire. Gravity, the classic force at a distance, does not release bowling balls once it has attracted them to the surface of the Earth. The wires used to conduct the electricity had been called "poles" by analogy with magnets. But Faraday said that the field description did not fit this case, and after consulting others, changed the name from pole to *electrode*. The equivalents of electricity, after 1874 and by suggestion of George Johnstone Stoney (1826–1911), became known as *electrons*.[2]

The next step was to take these electrons out of solution and into the air where they could be dealt with without interference. It is hard to say who first set up an electrical discharge through a partial vacuum. But by at least the late 1850s, someone had placed a negatively charged electrode (cathode) in a glass tube opposite a positively charged electrode and seen a discharge (ray) between them. Soon playing with cathode ray tubes became a favorite pastime of chemists and physicists, including Faraday and almost anyone who could blow glass and build a battery. By drilling holes in their electrodes, Eugene Goldstein (1850–1930) and his student Wilhelm Wien found the discharge actually consisted of two rays, one negatively charged and one positively charged.[3] They dubbed the positively charged rays canal rays because they went through the holes, or canals, of their electrode. A great deal of science has its origins in the study of both these rays.

The fascinating canal rays and cathode rays caught the interest of the English physicist Joseph John Thomson (1856–1940) and contributed to his earning the Nobel Prize in Physics in 1906.[4] Though he has been credited with "discovering" the electron, this is like saying Columbus discovered America: Plenty of other people knew it was there, they just weren't looking at it quite the way he did. Thomson took an intellectual risk. On April 30, 1897, J. J. Thomson declared that cathode rays were negatively charged particles with a mass about 1,000 times smaller than the hydrogen atom, the least massive of all elements. He went further to say that these corpuscles were building blocks for all matter. He proposed a model for the atom: a blob of positive charge with embedded electrons. This model, which has come to be known as the plum pudding model, did not contain protons as separate particles.

Many of the same dabblers, professional and amateur, who were fascinated with cathode rays were also fascinated with the new techniques of photography. Though several of the photographers knew and even shared the information that photographic plates kept too close to cathode ray tubes tended to go bad, no one stopped to investigate the cause until Wilhelm Conrad Röntgen.[5] The German physicist Röntgen (1845–1923), who received the first Nobel Prize in physics in 1901, took photographs of his own bones from the emanations of the cathode ray tube.[6]

When it was established that glowing cathode ray tubes emitted penetrating rays that Röntgen named X-rays, another question arose: Do other glowing bodies emit X-rays? Henri Becquerel (1852–1908) was well positioned to answer this question because Becquerel's father as well as Becquerel had studied minerals capable of phosphorescing (that is, glowing even in the absence of light).[7] Glow-in-the-dark materials require exposure to light before they start glowing. So at the start of his tests, Becquerel placed his minerals on photographic plates wrapped in black paper and set them in sunlight to start them glowing. One day the sun didn't shine, so Becquerel took the sample he had intended for study, a chunk of uranium, and stuck it in his drawer. After a few days, probably thinking that it would be good to have an experimental control, that is, a sample for comparison that hadn't been subjected to the stimulus, he decided to develop the film anyway. To his surprise, he found his control plate had been exposed, too. The uranium didn't need sunlight to activate it. It glowed with emanations of its own. For his discovery of this radioactivity (named by Marie Curie four years later), Becquerel was awarded the 1903 Nobel Prize for physics.

As more radioactive elements were found by the Curies and others, more scientists took up the problem of attempting to characterize these new "rays." Ernest Rutherford, born into a New Zealand immigrant family of twelve children, came to work in J. J. Thomson's

Fig. 20. Sir J. J. (Joseph John) Thomson, Edgar Fahs Smith Collection, University of Pennsylvania Library.

lab in England. After his initial exposure to cathode rays in Thomson's lab, he became caught up in the excitement over radiation. In his first studies of radioactivity, Rutherford accomplished three things: He established that the radiation from uranium and thorium was particulate, not electromagnetic radiation; it consisted of two different particles, which he named alpha and beta; and it was the result of atomic, not molecular disintegration, that is, the result of one element losing a piece of itself and turning into a new element—the first evidence that Newton had been right about transmutation. For this work Rutherford received the Nobel Prize in chemistry in 1908. But while these findings were important and required perseverance and talent to produce, if Rutherford hadn't found them, it is likely someone else would have done so shortly. That alpha particles are material particles and not electromagnetic radiation can be shown with paper: A piece of paper held in front of an alpha source will block the radiation. It was Rutherford's work on atomic structure, after he received the Nobel Prize for his characterization of alpha and beta particles, that was the true mark of his genius and for which he is most remembered.

The device that made Rutherford's work on atomic structure possible was invented by another devotee of the cathode rays, William Crookes (1832–1919), eldest son of sixteen children born to a London tailor, a student of the Royal College of Chemistry, and an acquaintance of Faraday.[8] The financial demands of his own family of ten children tended to encourage Crookes to apply his talents to enterprises and inventions as much as pure research. In one of his investigations of cathode rays he originated a device call a radiometer, a vane with black and white colored fins housed in an evacuated glass bulb that rotates when exposed to light, a device now sold as a toy.[9] The device that Crookes had invented in 1903 that Rutherford found so useful was a microscope that could be used to see the mark of alpha particles as they struck zinc sulfide. With Crookes's device, Rutherford could track the fate of alpha particles as they passed through thin slices of metal.

Rutherford's purpose was to confirm or, at best, refine Thomson's model. He expected that the alpha particles would encounter atoms of metal on their way through foils and be knocked at a bit of an angle. Most of the alpha particles went straight through without changing their course at all, a few were scattered at a large angle, but, to Rutherford's surprise, a very few rebounded directly back to the source. Rutherford scratched his head and did calculations for a couple of years before he committed himself to a new model of the atom: a very small, very dense positively charged nucleus, surrounded by a lot of empty space, and then an outer shell of electrons. The plum pudding turned into a little solar system. Scientists began searching for positively charged particles in the nucleus.

A young man named Henry Gwyn Jeffreys Moseley (1887–1915) joined Rutherford's group and asked to follow up on some rather tantalizing hints from the work of others.[10] He wanted to see if the X-ray spectra of elements might indicate how many protons were in the nucleus. At this point it should be remembered that appropriating a colleague's area of endeavor had earned Faraday the censure of his peers, but now in this time of rapid discovery it was anyone's game, as perhaps it should be. Anyone can propose that the moon is made of green cheese, but the Nobel Prize will go to the person who provides the evidence.

Prior to Moseley, a very talented and strong-willed chemist by the name of Mendeleev had posited that when elements are ordered according to their chemical properties, their order follows—roughly—their atomic mass. The word "roughly" is key because the relationship was not absolute. When ordered according to their chemical properties, on the *periodic table*, as it has come to be called, there are a few places that show a "mass inversion"; that is, instead of the steady increase in mass that is generally observed as the elements are listed, there are a few one-element dips. Such is the case with copper and nickel. Nickel comes after copper on the periodic table, but it weighs less. For Mendeleev this caused little concern. He was so certain of his chemistry that he shrugged and assured others that the mistake in mass would someday be found. After an amazing amount of work, Moseley finally determined that one of the X-ray lines in the spectra of elements changed in a predictable manner as he moved across the periodic table—including those elements that displayed a mass inversion. So Moseley and Mendeleev agreed: The mass of the elements did increase as one traveled across the periodic table, but the mass wasn't the factor that determined the position on the periodic table. What determined the position on the periodic table, Moseley found, was the magnitude of the positive charge in the nucleus, which was the number of protons.

Moseley was one of the few of Thomson's lineage who did not receive a Nobel Prize, but this was because of World War I rather than any lack on his part. Though Rutherford and others protested assigning trained scientists to front-line duty as a waste of their talents, Moseley refused the scientific position he was offered and was killed in action.[11] The Nobel Prize is not awarded posthumously.

After Moseley's work, a lingering question remained: Why didn't the mass increase as the number of protons increased? It really seemed logical that it should do so. When the answer came, it was from someone who was also affected by the war, though not as finally as Moseley.

Rutherford found new elements formed as a result of the atomic disintegrations that led to alpha and beta radiation. It fell to chemist Frederick Soddy (1877–1956) to investigate the properties of these new elements.[12] Soddy taught physical chemistry in England and in Canada, where he worked with Rutherford. In 1913, as a result of his research, he declared his finding that there could be atoms of the same element, that is elements with the same number of protons, that had slightly different mass. He called these different-mass atoms *isotopes*, meaning "same place" on the periodic table. Soddy became a social and political activist in the period between the world wars. His message was simple: He advocated responsibility to society from science. He became a socialist and ran for office in 1920. The attempt to bridge socialism and science, however, made him a stranger in both worlds, and consequently his political career fizzled. He did, however, find one friend: Herbert George Wells. H. G. Wells, as he is better known, wrote a novel, *The World Set Free*, based on an idea of Soddy's and dedicated it to him, though it is not one of Wells's better-known works.[13] In the book, a society manages to tap the energy of the atom through a new element called carolinium. They use this new energy source to build a utopian society that flourishes

Fig. 21. Frederick Soddy. Edgar Fahs Smith Collection, University of Pennsylvania Library.

until someone figures out how to build a bomb from carolinium. The ideal world then disintegrates in a war of total destruction. Published in early 1914, the book has never been very popular, perhaps hitting a bit too close to home.

Rutherford and others saw that Soddy's isotopes indicated that there must be another particle in the nucleus contributing to the mass but not contributing to the charge. The neutron, as this particle came to be known, was not found until 1932, when James Chadwick (1891–1974), Rutherford's student, found it.[14] He proved its existence by the effect it caused, à la Perrin. He measured the rebound of atoms the neutron hit to establish the mass of the object that hit them, and he won the 1935 Nobel Prize. With the confirmation of the neutron, the collection of the pieces of the atom was complete— but not the puzzle. To start fitting the pieces together, another of the descendants of Thomson, Niels Bohr, would have to return to spectroscopy. He would find his key in the strange pattern of lines found by Joseph Balmer in the spectrum of hydrogen.

Chapter 20
The Quantum Revolution

In this stage of affairs there appeared to us like a wonderful ray of light the beautiful hypothesis of [quanta]. . . . It has opened for us unexpected vistas, even those who consider it with a certain suspicion must admit its importance and fruitfulness.

Hendrik Lorentz, circa 1900

The phrase "quantum revolution" is a bit of a misnomer, though this is the traditional label applied to the accomplishments of Planck, Einstein, and Bohr in the early years of the 1900s.[1] Before Bohr could interpret Balmer's beautiful lines, a revolution certainly had to occur, but "quantum revolution" implies the concept of a quantum was revolutionary, which is not true. Dalton had already crossed that bridge. A *quantum* just means a very small portion. Dalton had proposed that matter comes in quanta called atoms, and he was backed by Boltzmann, Perrin, and Rutherford. What was revolutionary about the quantum revolution was the way science was forced to deal with the quantum world.

Quantum mechanics has its origins in thermodynamics, electro-dynamics, spectroscopy, and radiative heating. It may be recalled that the early thermodynamicists focused on conductive heating and neglected radiative heating because steam engines run as well in rain or sunshine. But there loomed in the background the question of the heat from the sun, and in the late 1800s, theoretical physicists, casting about for ways to make themselves more attractive as academic candidates, began to examine radiation on a more fundamental level.

The narrative begins with Gustav Kirchhoff, the wheelchair-bound physicist who had built the successful spectrometer with Robert Bunsen. Kirchhoff knew, as did others, that if the spectrum of the sun and the spectrum of a sodium flame were superimposed, some dark lines in the solar spectrum were replaced by the bright lines of the sodium flame. But Kirchhoff took the experiment one step further. He allowed the sun to shine though the sodium flame of the Bunsen burner before the sunlight was disbursed by the prism. When he did, he found these same dark lines in the solar spectrum got darker. He realized that this meant that the sodium in the flame was absorbing light from the sun—the same light that it could emit when heated.

Kirchhoff then asked the question, What if this were true of all materials? If all materials absorb light at the same wavelength that they emit, then it should be possible to construct a "blackbody," that is, a material that would absorb all frequencies of light (therefore be black), and, reciprocally, emit all wavelengths of light when heated.[2] If this blackbody were found, then it could provide the link between thermodynamics and spectroscopy: the link between heat and light.

Why was this a desirable goal? It doesn't take a Kirchhoff to know that hot bodies give off light. Turning on a stove burner will demonstrate this fact, and considerations of a stove burner will also point up the problem. As the burner warms, the wavelength of light it emits starts in the invisible infrared range and shifts toward the high-energy visible range, as can be demonstrated by putting a hand close to (but never on!) a warming burner. The burner will feel warm, that is, it will glow with infrared radiation, long before it glows with visible radiation, that is, before it turns red. However, it can be shown that even if the burner continues to heat, it will not continue to emit only higher and higher energy. If it did, it would eventually blind someone who looked at it. So even as a body gets hotter and hotter, the wavelength of light that carries the heat stays peaked in the visible . . . and the physicists didn't know why. According to classical physics, if more energy is added to the burner, then the light it gives off should be more energetic. Kirchhoff offered the following poser: Could an equation be found for the actual energy distribution of the light coming from a blackbody? This turned out to be a very good question.

In 1893, Wilhelm Wien (1864–1928), the student of Goldstein's who had worked on canal rays, devised an oven with a small hole to use as a blackbody.[3] The arrangement was a good simulation of a blackbody because any radiation impinging on the hole should be trapped inside (perfect absorber) and therefore, when heated, it should emit all frequencies (perfect emitter). In 1899 Otto Lummer (1860–1925) measured the output of this blackbody accurately and showed that the energy was distributed in a sort of skewed curve that peaked in the middle and died off at the end, not the exponentially increasing curve heading up toward infinity that classical theory predicted.[4] Much later, historians described the problem as the "ultraviolet catastrophe," but at the time it was not viewed as a catastrophe. It was a curiosity, but not one that was interfering with the train schedule, as we said. So Kirchhoff's poser remained until Max Planck tackled the problem.

Max Karl Ernst Ludwig Planck (1858–1947) did not have a particularly brilliant scholastic career and was not the first choice for Kirchhoff's position when Kirchhoff went to join Bunsen.[5] Before offering it to Planck, the Heidelberg University elders had offered the position to Boltzmann, who declined. For his research, Planck studied thermodynamics and applied himself to one problem in particular: showing that entropy could be explained by radiation and did not require Dalton's atoms for an explanation. He felt, as many did, that light, which you can see, is a better basis for building a theory than atoms, which you cannot. As such, he was familiar with the work of Boltzmann because he was trying to disprove Boltzmann. He was a good mathematician and had worked through all of Boltzmann's formulas for himself. So when Planck tackled Kirchhoff's challenge he had in mind the mathematics of statistical mechanics. He was aware that the Maxwell–Boltzmann treatment of the distribution of energies gave a skewed curve similar to the one he was trying to find, but their statistical analysis was built on the assumption of small particles of matter. As a first try Planck split his energies up into small packets so he could use the mathematics of statistics. The size of the packet, however, he made proportional to the wavelength of the light. The result was an equation that reproduced the skewed curve of blackbody radiation, but Planck didn't know why. He presented it anyway. The year was 1900 and Planck was forty-three years old.

As we now understand it, Planck's method of breaking energy into small packets worked because a blackbody gives off energy in packets, too. The molecules within a blackbody vibrate, which means the electrons and protons in the molecules are moving with respect to each other, which sets up electromagnetic waves, just like Hertz's spark. But the molecules can't vibrate at just any arbitrary frequency, they vibrate at particular set frequencies that are characteristic of the material of the blackbody. To excite a particular vibrational mode, just exactly the right amount of energy had to be absorbed, and no less. When a molecule vibrates, it gives off this amount of energy, a quantum of energy, and no more. Planck, however, did not make this intellectual quantum leap. He didn't know why his model worked.

There is nothing wrong with presenting a equation that works without necessarily knowing why it works. This approach has been used many times as a good first pass. For instance, Galileo could see that cannonballs flew in curves and derived equations to match that motion. Newton proposed the inverse square law for gravity because he knew that the orbits of the planets were ellipses. Planck chose to put energy into packets of increasing size because that created an equation that fit the data.

To explain why Planck's model worked to reproduce the skewed output of the blackbody, consider a field of cups in a rainstorm. Planck's packets of energy will be our cups, and as the size of Planck's packets depended on the wavelength of light, so will the size of our cups. Planck was considering the whole spectrum of light, which the-

Fig. 22. Max Planck. Edgar Fahs Smith Collection, University of Pennsylvania Library.

oretically stretches to infinity, but for practical purposes starts at low energy with radiowaves, has visible light toward the middle, and ends at high energy with gamma rays. For the sake of visualization, we will consider only visible light, which starts with red light at low energy and then proceeds through orange, then yellow, then green, then blue, then violet at the high-energy end. Because we want our cup size to depend on the wavelength, we will have our red cups be very short, our orange cups slightly taller, our yellow cups taller, then taller green, then taller blue, then have our violet cups be the tallest of all. Our cups have another property: When they fill with rain water they start to glow. Now let's look at what happens to our cups in a rainstorm.

We'll start with a gentle spring shower. All of the cups get wet, but only the smallest ones fill up. So the field glows with a gentle red, but that's all. When there is a heavier storm more cups fill up, including some of the bigger ones. The total glow from the field is stronger and is now a mixture of orange and red light—but no blue because there is not enough rain to fill any blue cups. With a still heavier rainstorm, the trend continues: The total glow from the field is even brighter, with contributions from the red, the orange, and the yellow. But even though the rainstorms may get progressively heavier, there still isn't any contribution from the tall cups. It just takes too much to fill them, and they have to be filled to start glowing. Even in a monsoon, the red, orange, yellow, and maybe even some green will be glowing and the total light may be quite bright, but there will be no contribution from the violet because it just takes too much water to fill the violet cup.

In our cup analogy, the rain corresponds to the energy input, or the heating of the blackbody. If each wavelength of light requires a certain minimum amount of energy to be put in before it is activated—if its cup has to be full before it contributes to the output—the output will be peaked in the middle and will not extend to the short wavelength end. Planck eventually became reconciled to his cup picture of the blackbody output, but he never liked it.

At first the equations of Planck were treated as interesting but not particularly useful because no one was sure why there should be different-sized energy cups in a blackbody. Then Einstein found another phenomena that could be explained by quanta. Einstein had always been interested in Maxwell's electromagnetic radiation, so he had studied Maxwell's equations and Hertz's experimental confirmation of them. Einstein was aware of an anomalous behavior Hertz had observed: That sparks generated at his detector could be made stronger by impinging light. Hertz had placed different light-blocking substances in front of the sparker and found that ultraviolet light was the only light that increased the spark. Hertz would not live to fully explore the effect but our friend J. J. Thomson would find that ultraviolet light caused electrons to be ejected from the metal and these extra electrons added to the spark. The effect, called the photoelectric effect, is used in some sensors such as the ones that open automatic doors. As you approach the door, you block or reflect a light beam and the alteration in the flow of electrons caused by that light beam can be detected by circuitry that opens the door. Philipp Lenard (1862–1946), another Nobel Prize winner, found that there was a certain threshold energy that was required to start the electron flow. In other words, the cup had to be full.[6]

In 1905, Einstein addressed the photoelectric effect. His paper on the photoelectric effect was one of three he wrote that year, his famous *annus mirabilis*, while working at the patent office, and, legend has it, while working in intellectual isolation without consultation with any other physicist. While this misconception has been perpetrated by many Einstein biographers,[7] recent biographers have begun telling the story a bit differently. He did consult with another physicist: his wife, Mileva Maric.[8] That Maric helped Einstein in his work is now fairly well established.[9] How much of a help was she to him? That is hard to say. Einstein had unquestionable genius and a wonderful ability to see connections where others could not. But his mathematics was not all that strong, as he would willingly admit, and thus throughout his career he sought help from mathematicians and other physicists with a more mathematical bent. It only makes sense that he would ask help from the person sitting across from him at the dinner table.

In his paper on the photoelectric effect, Einstein stated that the effect may be explained if one assumes light comes in small packets called quanta, little bits of light to go with Planck's little bits of energy, though Einstein only made brief reference to Planck's work. However, it is essential to note that Einstein was in no way rejecting Maxwell's equations or the wave model of light. What he was saying was that sometimes the wave model works well, and sometimes the particle model works well.

Many people since have found this concept difficult to assimilate. How can light be a particle sometimes and a wave at other times? The answer is that light is not a particle or a wave—sometimes it *acts* like a particle and sometimes it *acts* like a wave. Sometimes one model works best to describe the behavior, sometimes it's the other.

Both models work for sound waves, too. To understand how sound behaves as a wave, consider a wave passing through an opening. All waves tend to bend around the corner when they go through an opening (as water waves will bend around a pier), but the amount of bending depends on the size of the opening. If the size of the opening is about the size of the wave, the bending is pronounced. But if the size of the opening is many times bigger than the wave, then the wave "sees" open space and does not bend much at all. When a television is on in another room, you cannot see the television screen because light does not bend around doorways. The doorway is many million times bigger than the wavelength of visible light. However, you can *hear* the television because sound waves are big and *do* bend around corners.

To understand how sound behaves as a particle, consider a sound that can shatter glass. The sound has to be of just the right frequency, and when the glass shatters, it does so all at once, just as though it had been hit by a bullet. A bullet of sound. A particle of sound. This is what Einstein hoped to convey about the behavior of light, too: Sometimes it is best modeled as a wave, sometimes it is best modeled as a particle.

In 1907 Einstein used quanta of energy to explain anomalous heat capacities, that is, heat capacities that did not follow the behavior observed by Pierre Dulong and Alexis Petit. As it may be recalled, the heat capacities of solids were found to be a fairly constant function of their atomic mass; this in fact was taken as good evidence for the atomic theory. However, as cryogenic technology improved, it became apparent that the relationship was not holding at low temperature. At low temperature the heat capacity fell off drastically with

Fig. 23. Albert Einstein. Edgar Fahs Smith Collection, University of Pennsylvania Library.

temperature, that is, all heat seemed to go immediately into raising the temperature and none of it went into other modes, such as vibration or rotation. Einstein showed that this behavior could be explained if one assumed that the energy of the vibrational modes is quantized.

The analogy we might use here is with a frozen bolt versus a freely turning bolt. If a bolt is freely turning, then any amount of energy will turn the bolt. If the bolt is frozen, then a threshold amount of energy is required to break the bolt free before it can turn. Einstein said that the vibrational modes required a threshold amount of energy to get them vibrating (or using our previous analogy, the cup had to be filled). At high temperatures there are many vibrational modes available, so any amount of energy put in will find some vibrational mode to absorb it and the energy will partition itself between vibrational and translational modes. At low temperatures, however, the vibrational modes are "frozen" so energy put into the system will go into raising the temperature until it is high enough to break loose the vibrations. The quantum model worked well again.

The next use for quantum mechanics was found by Niels Bohr (1885–1962), the son of a Danish professor of physiology and a mother from a family of academics. After completing his degree in physics he went to England to study with J. J. Thomson, but he did not get along well with Thomson.[10] Soon he went to work for Ernest Rutherford instead. With Rutherford, he found himself trying to put together a picture of the atom, and they weren't the only ones. Perrin had suggested a solar-system–type model. Hantaro Nagaoka (1865–1950) of Japan had discarded Thomson's plum pudding model, saying that positive and negative electricity would not intermingle; he suggested instead an atom with a positive central charge and electrons arranged in Saturn-like rings around it.[11] However, neither of these models predicated a quantitative relationship that could be tested.

In 1913, two years after completing his doctoral dissertation, Niels Bohr wrote the paper that would earn him the Nobel Prize for 1922, a mere eleven years later. Why the rapid recognition? Bohr found something quantitative that could be related to the structure of the atom: the atomic spectra. Bohr assumed that electrons are located in orbits around the central nucleus, and that the absorption of a quantum of light of just the right energy promotes an electron from one orbit to another. Then with this model and Balmer's formula for the spacings between hydrogen lines, he could calculate a spectroscopic constant that agreed with the one measured for hydrogen. In addition, it predicted the spectrum that could be observed for the helium ion (helium missing one electron). This quantitative agreement between theory and experiment won him the praise of his fellow physicists, though it bothered Bohr that it was not more general. He also was not pleased by the fact that he had to postulate moving electrons. Why did Bohr's electrons have to be moving? Because of the experimental results of another wunderkind, Pieter Zeeman, who had shown in 1896 that spectral lines could be split by a magnetic field.[12]

Pieter Zeeman (1865–1943), like Bohr, was only thirty-seven when he won the Nobel Prize. Zeeman's demonstration of the interaction of the electrons in the atoms with a magnetic field meant that the electrons had to be moving (Faraday showed that a current produces a magnetic field, a static charge does not). However, if they were moving, then this led to another puzzlement. Hertz had demonstrated pretty convincingly that moving electrons generate electromagnetic radiation. If the electrons were giving off light, then they should be losing energy and spiraling into the nucleus. But it is common experience that materials don't continuously glow and spontaneously self-destruct. As we will see next, the problem would soon be resolved, but when it was, it would not be in a way that would please Kelvin: It would require another quantum leap. More bits and bugs.

Chapter 21

Quantum Riddle

I think that I can safely say that nobody understands quantum mechanics.

Richard Feynman, 1965 Nobel Prize, Physics

Before Niels Bohr's conundrum could be attacked, an entire new system of mechanics had to be created: quantum mechanics. In spirit, quantum mechanics shares the goal of classical mechanics. Quantum mechanics, just like classical mechanics, is a system of equations that relates forces and motions. But quantum mechanics has atoms and electrons as its objects, not cannonballs, and therein lies a world of difference. The utility of quantum mechanics is indisputable. Through quantum mechanics we can explain and predict spectra, elucidate the structure of molecules, and even predesign materials to meet specifications.[1] Yet the premises on which quantum mechanics is based are, to put it mildly, philosophically disturbing. Galileo, Newton, and Lavoisier conditioned humanity to expect a predictable, regular, and programmable world and to reject mysticism and magic. But, as we will see, in quantum mechanics we seem to regress to the unexplained and inexplicable. The acceptance of quantum mechanics feels like an act of faith. It requires a certain liberation of mind to appreciate quantum mechanics—and it required those of liberated mind to conceive it.

The story begins in 1924 in France with the doctoral thesis of thirty-two-year-old Louis de Broglie (1892–1987).[2] As may be recalled, the main difficulty with Bohr's model was the conflict with classical electrodynamics: A negatively charged particle in orbit around a positively charged nucleus should radiate energy and spiral into the nucleus. De Broglie no doubt agreed that a negatively charged particle would do this, but an electron wouldn't—because an electron is not a particle, it's a wave.

Prince Louis Victor Pierre Raymond duc de Broglie was born into a venerable French family that had earned titles for its members through military service to the state, including support of the monarchy during the French Revolution.[3] De Broglie earned his doctorate in physics rather late in life because he took his first degree in history. World War I also intruded. Conscripted, he staffed the wireless communications position at the Eiffel Tower for the whole of the war. When he returned to his studies, he remained interested in the physics of waves such as those that had made the wireless communication possible and performed experiments in an X-ray laboratory set up in the family mansion.[4] In his doctoral thesis, de Broglie reasoned thus: If light, which is a wave, has particle-like properties, then why should not matter, which is a particle, have wave-like properties? He took Einstein's famous equation for the energy–mass equivalence, $E = mc^2$, and compared this energy to the energy of light, which depends on the wavelength, and arrived at an equation that related the wavelength of light to a quantity with the same units as momentum, mass times velocity. If light, which has wavelength, also has momentum, he reasoned, then particles, which have momentum, should also have wavelengths. An interesting speculation, but was it true? His equation gave a means for predicting the wavelength of an electron at a given velocity, a number that someone could measure, so pretty soon someone did.

The experiment took place in the United States where Clinton Joseph Davisson (1881–1958), the son of a contract painter and a schoolteacher, accidentally found that when he bombarded nickel with electrons in a cathode ray tube, a few of the electrons bounced off with the same energy as the ones he fired in.[5] He measured the distribution of these electrons in space and found that it peaked in two spots, but was still fairly spread out. Then in April 1925 an accident intervened in Davisson's life again: The pure nickel sample he was using as his electron target got in the way of an exploding liquid-air bottle and he had to clean it by prolonged heating. This type of prolonged heating is now carried out routinely in modern surface-science labs to achieve the effect Davisson discovered by chance. The many tiny crystals that made up his nickel sample re-formed into a single crystal with a very regular surface. When his electrons bounced off the many tiny ridges in his regular crystal target, he found that they formed a regular pattern, similar to that formed when Young's light passed through his narrow slits. But this wasn't light: This was electrons. After learning more about de Broglie's equation, he calculated a wavelength that de Broglie predicted for electrons that had the same energy as his and calculated the diffraction pattern these waves should produce. The patterns matched.

But when de Broglie presented his dissertation, these measurements hadn't been made yet, and de Broglie nearly failed. Luckily, one of the examiners, Paul Langevin, advised de Broglie to send the dissertation to Einstein for comment. Einstein declared, "He has lifted a corner of the great veil."[6] Needless to say, the dissertation was accepted.

Why did Einstein believe de Broglie had explained so much? Because a wave, if it exactly fits its container, can form a standing wave. An example of a standing wave is the sound-wave tone that can be created by blowing across the top of a narrow-necked bottle. As anyone who has tried to generate a tone by blowing across the top of a bottle can attest, just the right angle and power are needed to elicit a tone. This is because the sound wave generated has to be just the right size to fit the bottle. When the sound wave is of just the right size to fit the bottle, it will reinforce itself as it hits the bottom of the bottle and bounces back. At other frequencies, the wave does not self-reinforce, and dies out. The self-reinforcing wave is a standing wave. The container for the electron wave was the orbit around the nucleus. If the electron wave were of just the right wavelength, if the end of a crest exactly met the beginning of a trough, then it would form a standing wave around the nucleus. This explained two things. First, electrons don't spiral into the nucleus because they are in stable standing waves. Second, only certain wavelengths fit around the nucleus, so only certain wave orbits are stable. To go from one wave orbit to another, just exactly one amount of energy would be required and no other—a quantum of energy.

But now there was another problem. Davisson shared the 1937 Nobel Prize with English citizen George Paget Thomson, the son of J. J. Thomson, because the younger Thomson had seen the same effect for electrons fired through an ultrathin metal sheet. J. J. Thomson had declared that electrons are corpuscles; now his son had declared that they are waves. So who was right, father or son?

The answer is both and neither. It turns out that matter can be modeled as a particle or as a wave, just like light. The description will depend on the measuring instrument and the point of view. This same type of thought was summarized (more eloquently and thoroughly) by Bohr in 1927, in a principle he called *complementarity*, which meant that the full description of the electron includes knowledge of its wave nature and its particle nature, too. So with this philosophy and de Broglie's discovery, quantum mechanics entered a new phase of understanding.

The first to fully exploit the wave model of matter in quantum mechanics was the Austrian physicist Erwin Schrödinger (1887–1961). Schrödinger seems to fit our picture of an ideal quantum mechanic in that he did not adapt well to the academic mold—his early work was not stellar and his committees were not always unanimous in their recommendations for his advancement. And he certainly was a free thinker. Schrödinger had many mistresses as well as a wife with whom he stayed married,[7] and at one point he lived comfortably with a mistress and his wife at the same time, while the mistress was carrying his child. This creative lifestyle was mirrored in Schrödinger's creative approach to quantum mechanics. Around 1925 Schrödinger derived an equation that produced functions describing electron waves around the nucleus.[8] These electron waves had energies that agreed with the positions of Johann Balmer's mysterious lines. But to have his wave equation work, Schrödinger had to introduce something nearly as mysterious: an imaginary number, the square root of minus one.

The tradition of calling the square root of minus one an imaginary number is unfortunate. There is nothing otherworldly about the square root of minus one. Our old friend the Italian mathematician Girolamo Cardano called this number "fictitious,"[9] but since that time imaginary numbers had found a respectable place in algebra. Why did this odd little number surface in Schrödinger's equation for the electron wave? Because it has precisely the behavior needed for Schrödinger's wave: If the square root of minus one is squared, it equals minus one, if it is raised to the fourth power it equals one, if it is raised to the sixth power it equals minus one again, and so on. The values for the powers keep cycling between one and minus one. Oscillating. A wave. The right behavior to fit de Broglie's model, but imaginary numbers in real systems can be philosophically unsettling. And there was more trouble ahead.

A German researcher by the name of Johannes Stark (1874–1957) found that in an external electric field, the individual Balmer lines split into several lines, similar to the way Pieter Zeeman had found they split in a magnetic field.[10] An explanation for this behavior came from the Dutch physicist Samuel Goudsmit (1900–1988). Goudsmit suggested that the electron was spinning while it was orbiting and generated a second magnetic field that could then interact with an electric field.[11] Fine. With one problem. That model went right back to the miniature solar system picture of the atom.

How can an electron be spinning if it is a wave? It can't . . . which means it's not a wave . . . but it's not a spinning top, either. Spin is another model for electron behavior, but the electron cannot be literally spinning because if it were it would be spinning faster than the speed of light. While many learned minds were discomfited by this fuzzy view of nature, then as now, there gradually arose a group of people who were not so disturbed. One in particular was the German Werner Heisenberg.

When twenty-four-year-old Werner Heisenberg (1901–1967), research assistant to Bohr, stood for oral exams he nearly failed. One of the professors asked him a laboratory question related to the resolving power of a microscope, and Heisenberg got it wrong. He still got it wrong when he proposed his famous uncertainty principle two years later,[12] but it was a minor correction on a major idea.

Basically what Heisenberg said was that there was no use in trying to fit the atom into the classical mold because there was no way of measuring the atom for the fit. To apply classical mechanics to electrons,

Fig. 24. From left to right: Quantum physicist Victor Frederick Weisskopf, Maria Goeppert Mayer (see Epilogue), and Max Born. AIP Emilio Segrè Visual Archives.

one would have to know where the electrons were and how fast they were moving. But given the size of the electrons and our optical instrumentation, Heisenberg said this would never be possible. Actually the statement is stronger than that: It is not just a limitation of our instrumentation, it is a limitation of nature.

The problem is that a light wave will knock an electron head over heels—hence no simultaneous measurement of position and velocity is possible. It would be like trying to measure the position of a ping-pong ball by bouncing a bowling ball off it. Heisenberg, of course, stated this principle more mathematically and eloquently and made it more general, but the root cause of the problem is the same.

Many were not happy with the new turn of events, notably Einstein and even Schrödinger, but others were more able to live within the limitations Heisenberg imposed. One in particular found a creative rationalization, German Max Born, the originator of the descriptor "quantum mechanics."[13] Max Born (1882–1970) likewise had a lackluster scholastic career and originally intended to study history, but took up science to please his father.[14] Born proposed that even if you could not throw a saddle on the electron, you could still ride the wave: He interpreted Schrödinger's waves as probability waves. They did not describe where the electron was, but where the probability for finding the electron was the highest.

With this pronouncement, Born sounded the final death knell for the predictable, neat, geometric universe of Aristotle, Newton, and Galileo. Granted, Boltzmann and Maxwell had used statistics to describe their system of particles. But there was always the underlying assumption that if you cooled down these particles, or trapped them in a field, they would again start acting as well-behaved billiard balls and would stand still long enough to be measured. But now Heisenberg said that this could never be so: An electron never stands still.

Heisenberg's uncertainty principle and Born's probability wave became known as the Copenhagen interpretation of quantum mechanics, named for the school of thought that grew up around Neils Bohr at the University of Copenhagen in the capital of Denmark.[15] As one would suspect, these theories produced considerable debate, some of it bordering on antagonism.

But it was an antagonistic age. Of the German scientists, Einstein, Bohr, and Born eventually had to flee the Nazi regime and the Austrian Schrödinger was persecuted into emigration when he spoke out against the Nazi government. Heisenberg remained in Germany but was called a "white Jew" by Stark and Zeeman because of his lack of enthusiasm for the Nazi government. To this day there remains speculation that he purposefully delayed the development of the German atomic bomb, though there can only be as much uncertainty in this as he envisioned for the electron.[16]

Despite the difficulties, political, personal, philosophical, and otherwise, the work of these freewheelers endured. The sentiment was summarized in 1929 by Paul Dirac:

> The general theory of quantum mechanics is now almost complete, the imperfections that still remain . . . are . . . of no importance in the consideration of . . . ordinary chemical reactions. . . . The underlying physical laws necessary for the mathematical theory of a large part of physics and the whole of chemistry are thus completely known, and the difficulty is only that . . . these laws lead to equations much too complicated to be soluble. It therefore becomes desirable that approximated practical methods of applying quantum mechanics should be developed, which can lead to an explanation of the main features of complex atomic systems without too much computation.[17]

Finding "practical methods of applying quantum mechanics" is exactly what the physical chemists proceeded to do.

Part V

The Flourish of the Physical Chemist

INTRODUCTION

This is physical chemistry, formerly a colony, now a great free land.
Jacobus van't Hoff, circa 1900

One definite outcome of the new theories in physics—from electrodynamics to quantum mechanics—was the emergence of scientists both versed in physics and grounded in chemistry. The new work in the overlap areas between chemistry, physics, and mathematics was sometimes called *chemical philosophy* or *theoretical chemistry*, but by the late 1800s the name *physical chemistry* came into general acceptance. In the new era, many of our heroes are now born-and-bred physical chemists, though contributions continue to come from those with roots in other disciplines: mathematicians with a background in physics, physicists with a background in chemistry, and engineers with a background in all three.

The major areas that lacked a firm theoretical footing at the beginning of the 1900s were chemical bonding, equilibrium, and the speed with which a reaction occurs. Given all the wonderful progress in physical theory, why couldn't the physicists derive a set of equations that would describe bonds or mimic the behavior of chemicals

247

in reactions? Basically, because atoms are not billiard balls and doing chemistry is not playing pool. To begin with, atoms and molecules are sticky. They don't rebound perfectly from each other and the walls. And they aren't always electrically neutral: Charged particles, *ions,* have a chemistry all their own. And molecules don't only exist as gases. They become liquids, solids, and complicated mixtures and solutions, and then have the audacity to bounce around from one phase to another. Add to this, every reaction involves an interface of some sort, even if it's only between solution and container. These factors provide a fairly clear picture of what the physical chemists were—and are—up against.

In response to the complexity and enormity of the task, there arose two schools: the Ionists and the quantum chemists. The Ionists took the top-down approach. They applied the methods of physics to describe the behavior of chemical systems. A skeptical lot, they confined their speculations to effects that could be quantified by their careful measurements. Unwilling at first to even admit to atoms, they developed powerful empirical relationships that set the bar for the theorists. The quantum chemists approached chemistry from the bottom up. They used the power of quantum mechanics and mathematical models to discern the structure of molecules and from this structure attempted to predict how chemical systems should behave.

These two approaches were, to some extent, geographically separated as well. The development of quantum chemistry took place primarily in the United States, while the Ionists thrived in Europe, especially Germany. This division, of course, was not absolute. Many of the physical chemists in the United States had been educated in Europe and the European scientists often visited the United States on lecture circuits and to confer.

In this part of our story we will trace the development of quantum chemistry and see how it merged with the observations of the Ionists. These physical chemists will deal with the challenging questions of intermolecular forces, the three-dimensional nature of molecules, and the statistical nature of energy and entropy. But being challenged is not the same as being confounded. The physical chemists would persevere.

Chapter 22

Rebels and Radicals

Planck . . . and Bohr . . . have invented systems containing electrons of which the motion produces no effect upon external charges. . . . [N]ot only [is this] inconsistent with the accepted laws of electromagnetism, but I may add, is logically objectionable, for that state of motion which produces no physical effect whatsoever may better be called a state of rest.

G. N. Lewis, physical chemist, 1916

Initially the premises of quantum mechanics ran somewhat counter to the chemist's experience and intuitions. By applying the chemist's insights to the physicist's models, a link between quantum physics and quantum chemistry would eventually be forged, though the fit, albeit quite serviceable, would never be exact. The task required those with the insight to see connections where others could not. The United States, unique in its time for being a country born of revolution rather than tradition,[1] and populated by people who traced their heritage to refugees, outlaws, opportunists, slaves, and nomads, served as a breeding ground for such liberated spirits. Add to that some of the most talented scientists in the world fleeing Hitler and war-torn Europe, and the results would be remarkable.

The first problem confronting both chemists and physicists intent on applying quantum mechanics to chemistry was to explain the chemical bond. This was no trivial task. They knew that the main force holding atoms together in molecules had to be electrical. Humphry Davy and Michael Faraday had offered ample evidence that electricity disrupts bonds in salts and forms them in metals. And electrostatic interaction was well understood. The strength of the attraction between positive and negative charges is called Coulomb attraction, and the fact that the force falls off with the square of distance, like gravity, is called Coulomb's law—though Coulomb was not the first to discover Coulomb's law. Two of the earlier discoverers kept their finding to themselves in the protective style of the medieval mathematicians, another published in Russian, and another presented his results only at a meeting. In 1785 Charles Coulomb did a crude experiment, published immediately, and was immortalized.

We used adhesive tape to show how electrical repulsion varies with distance and we can use adhesive tape to demonstrate Coulomb attraction, too. Again a strip of tape is pulled vigorously from the roll and attached to the side of a table so that most of the strip dangles free. If this time the remaining roll (or any other good source of electrons) is brought close to the dangling strip, it will attract the strip, and the attraction will rapidly becoming stronger at shorter separations.

But while electrostatic attraction explained the bonding in salts such as sodium chloride (table salt), where the sodium ion and the chloride ion have opposite charges, it did not explain why two atoms of the same element would be attracted to each other. We mentioned previously that hydrogen gas is formed from two atoms of hydrogen; oxygen gas is formed from two atoms of oxygen; and nitrogen gas is formed from two atoms of nitrogen; but neither the chemists nor the physicists could say why. The chemists and physicists also could not explain the observed *valence*, or combining power of elements. For instance, it was well established that carbon likes to form attachments to four other entities in hydrocarbons, therefore has a valence of four, while hydrogen attaches only to one thing at a time, and has a valence of one.

The path to understanding began with the results of Henry Moseley and Dmitri Mendeleev. Moseley showed that the number of protons in the nucleus, and hence the number of electrons in a neutral atom, increases regularly across the periodic table. At the turn of the century, a new group of elements were discovered: helium, neon, argon, krypton, xenon, and radon. The first one, helium, was found in the sun by the spectroscopist Joseph Lockyer, but it took longer to find the rest because these elements are singularly unreactive. For instance, we would never consider giving a hydrogen balloon to a child to play with because hydrogen is too reactive, too explosive. But we give helium balloons to children to play with all the time because helium is unreactive. As these elements were discovered, some argument arose as to where to place them on the periodic table. Most people agreed that they belonged in their own new column to be tacked onto the right side of the table. Mendeleev protested because he was ready to retire and was not interested in revising his table, but in the end he was overruled.

When they assumed their rightful place, a peculiar pattern emerged: helium, which finished out the first row, had two electrons, but neon and argon, which finished out the second, had ten and eighteen, respectively. Two for helium plus eight to get neon plus eight more to get argon. And these elements were particularly stable. Was eight a special number that gave stability?

In England, J. J. Thomson, who was familiar with chemical reactivity, proposed the electrons arranged themselves in shells in his plum pudding model. The physical chemist Charles Rugeley Bury (1890–1968), a veteran of an Iranian campaign in World War I, used chemical evidence to propose that electrons arranged themselves in layers of two, eight, and eighteen, which has subsequently turned out to be correct.[2] Credit for this idea is often given exclusively to Neils Bohr, perhaps due to Bohr's penchant for enthusiastically embracing ideas without always taking the time to acknowledge the source.[3] Thomson's and Bury's observation, along with much experience with chemistry, led German chemistry professor Richard Abegg (1869–1910) to suggest that the basis of chemical reactions is elements seeking to have eight electrons around each nucleus.[4] They achieve this, posited Abegg, by sharing electrons with other elements in close contact. However, Abegg was killed in a balloon accident when he was just forty-one and did not have time to explore his idea fully. The chemist who did explore this idea in depth would hail from the United States.

Gilbert Newton Lewis (1875–1946), could trace his lineage to Plymouth Colony of 1632.[5] Lewis (or G. N. Lewis, as he is affectionately referred to by the chemical community) could read by the

Fig. 25. G. N. Lewis. Courtesy of the Bancroft Library, University of California, Berkeley.

age of three and reportedly read *Robinson Crusoe* by the age of five.[6] He was tutored at home until his family moved to Lincoln, Nebraska, at which time he entered a university preparatory school and then entered the University of Nebraska. He transferred to Harvard to complete his schooling, where he earned a Ph.D. in physical chemistry by age twenty-four. After a year of teaching he performed the requisite pilgrimage to study at the premier institutions in Europe and visited the labs of some of the Ionists. When he returned, conflicts with his research advisor at Harvard, T. W. Richards, a Nobel Prize–winning chemist, inspired Lewis to serve for one year in the Philippines as superintendent of the Bureau of Weights and Measures. This was a brutal, ignominious episode in U.S.–Philippines history, and Lewis did not often speak of his experiences there. But he did manage to publish in that year and when he returned to the United States he joined the faculty at the Massachusetts Institute of Technology (MIT), though he was lured away seven years later to Berkeley. Before he moved west he married Mary Sheldon, the daughter of a Harvard professor.

He stayed at Berkeley for some thirty-four years and published papers on topics as far-ranging as relativity, isotopes, thermodynamics, and the role of unpaired electrons, or *radicals* as they have become called, in photochemistry. In fact, the bulk of Lewis's work was more in line with the efforts of the Ionists than the quantum chemists and will be considered in that context. His contribution to quantum chemistry—the two-electron bond—is the concept we examine here.

Apparently the idea occurred to him when he was teaching at Harvard. While sketching the arrangements of electrons around the nucleus for various elements, Lewis mistakenly drew helium with eight electrons because he knew helium to be exceptionally stable. He caught his mistake, but it started him thinking about the number eight and what it might represent. He took to drawing the electrons around a nucleus at the eight corners of a cube (perhaps to help himself keep track of the electrons and not make further mistakes at the board). He then found that by allowing incomplete cubes to fill in their missing parts by sharing sides with other cubes, that is, linking the cubes together like puzzle pieces, he could build compounds on the chalkboard that had all the corners of their cubes filled. When he stood back and looked, the resulting figures had ratios of elements that were the same as known formulas for compounds. Lewis had found a rationale for a bond without having to invoke electrostatic attraction: Bonds formed by sharing sides, by sharing pairs of electrons, so that their eight-cornered cubes were complete. He also had found a rationale for valence, the number of attachments an atom is apt to take on: Cubes sought the number of other cubes needed to complete their sides.

Did atoms stack together like building blocks to form molecules or was this just a fortuitous device? The idea was shaky enough that the other faculty at Harvard did not stand up and cheer (in fact, his research director, Richards, denigrated the "Twaddle about bonds: A very crude method of representing certain known facts about chemical reactions")[7] and Lewis, sufficiently dissuaded, did not publish his idea. But he stored it away in his mind.

The biggest problem with the cubic model is that it requires static electrons. By then Pieter Zeeman in Germany had shown that spectral lines would be split by a magnetic field, which indicated that electrons in atoms were moving. But chemists knew that the structures of materials such as crystals were rigid, which was difficult to explain in terms of a moving electron. Later, at MIT, when Lewis saw that others were proposing ideas similar to his, he decided to come forward with his cubes. Lewis showed how electrons completing the cube explained the charge carried by common ions and used his cubic atom to explain the structures of a variety of compounds that had not been previously explained. He also postulated an explanation for nonelectrostatic bonding: The two electrons of the bond, the two electrons that formed the edge of one cube, were spinning and therefore creating a magnetic field. If the two electrons had opposite spins, then their magnetic fields would be opposite and attract. This idea was not retained in future bonding theories, but it worthy of note that Lewis provided an early, intuitive, postulation for electron spin in the atom.[8] At the time, however, his theory received no particular notice.

In 1919, a chemist with a broader reputation, Irving Langmuir (1881–1957), adopted and extended Lewis's ideas and began lecturing on them, though always giving credit to Lewis as the originator.[9] Lewis, ever sensitive to slights, real or imagined (he refused an honorary degree from Harvard, still smarting from the treatment he received there as junior faculty), exchanged a series of restrained but pointed letters with Langmuir, though a warmer relationship eventually developed.

The two-electron bond described by Lewis was picked up by the German physicists Walter Heitler (1904–1981) and Fritz London (1900–1954) in 1927 when they tackled the problem of finding a model of the chemical bond that incorporated quantum mechanics.[10] They were glad to have a starting premise, although they reasonably chose the simplest possible bond to start with—the bond between two hydrogen atoms. They knew they were going to end up with an approximation, so they wanted to start as close to reality as possible. How did they know that they would end up with an approximation? Because even helium, with just one nucleus and two electrons, had yet to be solved exactly. For helium the problem is called the three-body problem: Interactions between the positive nucleus and both negative electrons have to be accounted for as well as the repulsions between the electrons themselves. For the hydrogen molecule the problem is even more complicated: The hydrogen molecule has two nuclei and two electrons, and they are all moving around with respect to each other.

Heitler and London considered electrons on atoms to be moving in the fuzzy electron orbits (electron clouds) described by the Copenhagen interpretation of quantum mechanics. They envisioned a chemical bond forming by a sort of coalescence of electron clouds: somewhat like the way two bubbles on the surface of a liquid might merge into one. To model this behavior, they first simplified the picture by assuming that the motion of the electrons is so fast relative to the nuclei that the two nuclei can be considered to be standing still, an assumption called the Born–Oppenheimer approximation. Next they turned to the mathematics of statistics[11] and said that the probability of finding two electrons together at any point in space is the product of their separate probabilities. The well-known result that probabilities should multiply can be understood by considering two coins. The probability of flipping one and have it come up heads is one out of two, or one half. The probability of flipping the other and getting heads has to be the same. But the probability of flipping both and getting two heads is less likely (and can be satisfactorily demonstrated by a series of 10 or more duo coin flips), which is the case for multiplied probabilities: One half times one half is one fourth. According to the so-called Copenhagen interpretation, the probability for finding the electron is given by the function for the electron wave. The imaginative leap that Heitler and London made was to assume that electron-wave probabilities behaved like coin-flip probabilities. They treated the total probability for finding an electron on two nuclei as the product of the probabilities on the separate nuclei; that is, they multiplied the two probabilities together by multiplying the two wave functions together.

Of course electrons are not coins, so some adjustments had to be made for the realities of the electron world. Werner Heisenberg had pointed out that because the electrons are indistinguishable (you can't put a label on something you can't pin down) any complete solution would have to reflect this blurriness. To account for this restriction, Heitler and London added on the product of the wave functions for the hydrogen atoms with the electrons exchanged. Heitler and London then put the composite wave function in Erwin Schrödinger's wave equation to find the total energy. They found terms for the electrostatic attraction between electrons and nuclei; the repulsion between the two nuclei and the two electrons; and, in addition, an odd "exchange" energy that results from quantum fuzziness. This fuzziness derives from the fact that you can't find the position and the velocity of an electron at the same time, which was Heisenberg's uncertainty principle. There is no classical analog for this exchange energy, but the exchange energy, as it turns out, is much greater than the electrostatic energies. The magnitude of the exchange energy shows that quantum mechanics has to be evoked to make any quantitative predictions concerning the chemical bond, such as the binding energy (the energy required to break the bond). However, using the Heitler–London model, the calculated values for the bond energy of the hydrogen molecule compared well with experimental values. The Heitler–London theory became known as the *valence bond theory* because it combined the waves of electrons located on separate nuclei to form a bond, in essence treating the hydrogen molecule as an atom combining with another atom.

Heitler and London were both forced from Germany by the Nazis, Heitler for his Jewish ancestry and London for being a Jew. Heitler stayed in Europe, although he was interned as an enemy alien in Britain for a short time because of his German citizenship. He eventually settled in Switzerland. London settled in the United States, teaching at Duke University in Durham, North Carolina.

But though the work of Heitler and London was progress, there is very little chemistry that involves only hydrogen gas, so a more general approach was needed. The torch was taken up again in the United States, and this time by Linus Pauling.

Linus Pauling (1901–1994) was born in Oregon. His father was the local druggist, which meant he sometimes served as a doctor, too, but he had a difficult time making much of a living at either. When Pauling was four, the family was forced to seek financial assistance from his mother's family, a circumstance that seems to have left a strong impression on Pauling's mother.[12] When Pauling's father died, his mother thought that Pauling should quit school and secure work to help his family. Pauling never did quite quit school, but he did have to work all the time he attended classes. In fact, he didn't formally complete his requirements for a high school diploma, but his high school graciously awarded him one after he won his second Nobel Prize. Even without the high school diploma, however, he was able to enroll in the chemical engineering program at Oregon Agricultural College, which later became Oregon State University. His performance at the college was such that he was given the position of instructor in quantitative chemical analysis, even though he had just finished the course. His future wife, Ava Helen Miller, was a student in this class.

Though he still had to provide substantial financial help to his family, he decided to go to graduate school, again while working. He wanted to study under Lewis at Berkeley but he couldn't wait around for the school to decide if they would accept him, so he went to California Institute of Technology (Cal Tech) instead, where he could work as a teaching assistant while taking classes. He received his Ph.D. there at the age of twenty-four.

A year later he received a scholarship to study in Germany, where he started working on the quantum mechanics of bonding. He ran into his own quarrels with notables over priority of ideas and was rather harshly accused of ethical malfeasance by British physicist W. L. Bragg, whom we will meet in another context, though Pauling had no difficulty speaking in his own defense. Then he was offered an assistant professorship back in the United States at Cal Tech. He believed he was receiving a joint appointment as professor of theoretical chemistry and mathematical physics, but when he arrived his plaque read "professor of chemistry." Miffed but undaunted, he proceeded to produce landmark work in theoretical chemistry and mathematical physics.

There were other theories of the chemical bond that were emerging at about this same time, but Pauling preferred to stay with modifications of the valence bond theory because it was something that could be visualized and made pretty good predictions for chemists.

To understand why the valence bond is something you can visualize, for a moment consider a rodeo and a cowherder doing rope tricks. In a typical rope trick, a cowherder twirls a rope, being careful to flip the rope, to put energy into the rope, at just the right time so that the rope maintains a stable orbit, tracing a figure such as a circle or a figure eight. An electron in its circuit around the nucleus can be thought of as doing something similar: tracing out a stable orbit that can be calculated and predicted by quantum mechanics.

One important difference between a rope trick and an electron orbit is that the orbits are three-dimensional instead of flat: The circle of the rope trick becomes a spherical orbit and the figure eight of the rope trick becomes a set of three three-dimensional dumbbell shapes. The dumbbell shapes appear to line up on an atom like the staff and arms of a weather vane: one arm pointing north/south, one arm pointing east/west, and the staff pointing up/down. But please bear in mind that the atom neither knows nor cares what we call north or south. In the absence of an electric or magnetic field, a collection of atoms would have these weather vanes oriented in all possible random directions. By visualizing orbits this way, one can speak of the bond that forms when these orbits overlap and begin to share electrons. When two electron-orbit weather vanes come close enough to touch, there can be an overlap and blending of the orbits to make a bond. There is one small problem, though. For even the simple molecule of methane, the orbits have the wrong shape.

A molecule of methane, swamp gas, is formed from a central carbon nucleus surrounded by four hydrogen nuclei. Because the electrons associated with the bonded hydrogens are all negatively charged, they can be thought of as repelling each other. Each of the hydrogen nuclei moves as far way from the other as possible, which results in a four-pointed figure, sort of like four balloons tied together by their ends, pointing in four opposing directions. But the best electron orbits that the carbon atom has to offer are the dumbbell orbits, which are pointed in six directions, not four. Pauling's inspiration was to mathematically combine the spherical orbit with the three dumbbell orbits (sort of a championship rope trick), which produces four new orbits that have the right directionality. This imbued valence bond theory with new life.

Pauling continued to develop the valence bond theory and reap the rewards of its application, but like everyone else, his work was interrupted by World War II. Though he did work on projects in support of the war, he turned down work on the Manhattan Project. He and his wife fought the internment of Japanese Americans, some of whom they knew personally. As he became more politically involved during and after the war he found himself beset by difficulties from the government. Not given to conciliatory sentiments, he took it on the chin more than a few times, despite his later accomplishments concerning biologically important compounds.

For the record, there were other people working on ideas similar to Pauling's, notably John Slater (1900–1976).[13] Like Pauling, Slater also recognized that atomic electron orbits could be combined to create new solutions to the Schrödinger equation and provide the three-dimensional features required for real-world molecules. Slater introduced the term *directed valence* for what he and Pauling were doing. British chemists, notably Nevil V. Sidgwick, predicted the shapes of molecules by employing Pauling's and Slater's ideas about the ability of electrons to adjust their orbits and positions to minimize repulsions.[14] The resulting theory, valence shell electron pair repulsion theory, was highly successful, predicting structures that could be verified by spectroscopy. These methods and theories, as well as ways of depicting atoms as a nucleus surrounded by static electrons, came to be called Lewis dot structures, and are still used today because of their power to predict three-dimensional shape and bonding. The valence bond approach is also still widely employed to visualize how atoms and molecules come together to form new compounds. Chemistry occurs when orbits overlap, and these simple models have impressively accurate predictive power.

But the chemists were aware that there had to be a more complete picture to bonding. The valence bond approach could be likened to a stick-figure model of the human form: very useful in some applications, but still a far cry from the real thing. There were unsolved problems in spectroscopy, statistical mechanics, and reactivity that needed a better model. But, as we have seen before, as better models emerge, the pictures become less simple, which is what happened in this case, too. But though some visual power was lost, the elegance of the system was certainly preserved. The new bonding model grew, once more, out of the work of the marvelous mathematicians.

Chapter 23

System and Symmetry

I heave the basketball; I know it sails in a parabola, exhibiting perfect symmetry, which is interrupted by the basket. It's funny, but it is always interrupted by the basket.

Michael Jordan, basketball player, circa 1990

To flesh out our stick figure picture of molecular orbits, to advance the art of divining molecular orbits from atomic orbits, more methods had to be imported from the mathematicians and from more mathematicians than there is room to mention. However, to pay our respects, we will acknowledge two that are particularly worthy of note: Emmy Noether and Evariste Galois. Although their work was in abstract fields of mathematics—symmetry and group theory—it will be these types of techniques that will be needed. What Noether noticed was that from snowflakes to rainbows to starfish to antimatter, there is an inherent connection between symmetry and natural law. What Galois noticed was that if you turn a starfish over, you still have a starfish, but if you slice it down the middle, you have a different beast.

Emmy Noether (1882–1935) was born to a Jewish–German family, the oldest of four children. Her father was a well-respected professor of mathematics. At that time in Germany, women had to get explicit permission from the professor to take a university course, but Noether's father's influence may have helped her. She earned her doctorate by the age of twenty-five, summa cum laude. Though she was not allowed to start a university teaching career, she continued to publish and live in the enriched academic environment of her father and his colleagues.[1]

In 1915, when she was thirty-three, she was persuaded by the faculty at Göttingen to teach there unofficially while they worked to have her added to the permanent faculty. In the arguments over her appointment, one of her supporters is said to have stated, "I do not see that the sex of a candidate is an argument against her admission as Privatdocent. After all, we are a university not a bathing establishment."[2] While at Göttingen, Noether gained an excellent reputation as a mathematician in the field of invariants; that is, mathematical objects that are not changed under certain transformations (such as turning the starfish over). Because of her skill, she was asked to examine a particularly sticky problem in the developing theory of general relativity. She found the solution requested and in the process developed a general theorem to cover similar situations. The essence of Noether's theorem was to show that symmetry is of fundamental importance in the physical world. Noether's theorem was read and praised by Einstein and others for its far-reaching applicability. Of course, praise from Einstein did her no good—or worse—when the Nazis came to power. Bryn Mawr College in Pennsylvania was nonetheless glad to receive her, so she moved to the United States. She taught there and also at the Princeton Institute for Advanced Study until her death at the age of fifty-three.

Noether's theorem, though more far-reaching than is being presented here, can be explained by a very practical concern: surgery. There is an inherent symmetry in the human body and any operation that disturbs that symmetry is not good medicine. Someone going in for a nose job does not want to come out with an ear attached to a leg. Similarly, any operation applied to a model of a physical system, such as a wave function, cannot change the symmetry of that model if any believable result is to be obtained. Applying Schrödinger's equation to ethanol, which has oxygen attached to carbon and hydrogen, should not return the energy for diethyl ether, which has the same atoms but with oxygen attached to two carbons. So how does one know if an operation is going to disturb the symmetry? Read on.

Our second hero has a sad story. Evariste Galois (1811–1832), was born in France and was an indifferent student until he took his first class in advanced mathematics at the age of sixteen.[3] During his final university exams, his mathematics examiner was impressed with his knowledge, but his language examiner was worried, believing the boy might even be half-witted. Once out in the professional world, he was encouraged by respected mathematicians of his day, but bad luck haunted him. He submitted an important paper to be considered for a prize, but the paper was misplaced when a referee died. He tried the well-trod path of tutoring for a living, but his communication skills were so poor that no one would pay him his fee.

The famous mathematician Sophie Germain knew of his plight and wrote a letter to a friend describing his financial and emotional difficulties, but by that time he was beyond help, even if it were forthcoming. He was arrested for making drunken threats against the king, acquitted, and arrested again. On his release, he was killed in a duel over love and politics, though the exact reasons were never clear. Abandoned by his second, he was found dying by a peasant. His brother and a friend, following his wishes, copied his mathematical papers (some of which, tradition has it, were scribbled on the last night of his life) and sent them to famous French mathematicians. Finally, in 1846, fourteen years after his death, Galois's work was understood, fully appreciated, and published.

What Galois had discovered was the basis for what became *group theory*: for any given solution of an equation there is a group of operations that are associated with it. What does this mean for physical chemistry? It means that when the symmetry of a molecule is known, there is a group of operations that will preserve the symmetry of the molecule

Fig. 26. Emmy Noether. Courtesy of the Bryn Mawr College Library.

and other operations that will not. Group theory provides us with the set of mathematical operations that preserves the symmetry of the molecule: just as invariant as you would wish to be after a surgical procedure or upon seeing your reflection in a mirror. This is as important a consideration in physical chemistry as it is personally: Molecules are changed by chemistry, not mathematics. The first important application of symmetry and group theory to consider will be the development of the prime rival to valence bond theory: molecular orbital theory.

In 1929 John Edward Lennard-Jones (1894–1954) suggested deriving functions for molecular orbits by adding atomic orbits instead of multiplying them.[4] The result was one big orbit around the bonded nuclei: a molecular orbit for the electron. But how did one go about choosing which atomic orbits to add? The spheres? The dumbbells? The complicated rope tricks that emerge after that? And how would you know when to quit? Almost any arbitrary shape can be created from a set of building blocks if you use enough of them. The answer was symmetry. Once the symmetry of the molecule was fixed, Galois's work suggested that there would be a limited group of operations that could manipulate the molecule and leave the symmetry unchanged. So any atomic orbit that contributed to the molecular orbit would also have to survive these operations with its symmetry unchanged. These restrictions narrowed the field enough to make the calculations tractable. The resulting wave function could then be put into the Schrödinger equation and solved for energy.

But these new molecular orbits manufactured from atomic orbits lost their resemblance to their original structures. The sacrificed spherical and figure-eight orbits were replaced by odd-shaped composite clouds that extended over the entire molecule. The word *orbital* was introduced by R. S. Mulliken to stress that these electron clouds were regions of electron probability, not electrons circling a nucleus (Heisenberg had shown that you could never locate an electron long enough to know if it were moving in an orbit). Mulliken received the Nobel Prize in 1966 for his molecular orbital theory.

Naturally a question arises: Why work so hard on a new bonding theory when you already have one that seems to be functional? Initially, at least, this was a nontrivial question and there were some hostile words exchanged between those of the valence bond camp and those of the molecular orbital camp. Both theories produced energies that were slightly off the mark. It could be argued that time would be better invested in improving valence bond theory than clouding the issue with a new method. But these arguments fell by the wayside as profitable applications of molecular orbital theory continued to surface. For one, molecular orbital theory had some interesting things to say about reactivity.

A cloud really isn't a bad image for a molecular orbital. The molecular orbital clouds are a bit more uniform and smoother than the ones we are used to seeing in the sky, but the lumps, bumps, and extensions into space of molecular orbitals are cloud-like. If idealized, smooth clouds can be imagined, this would be close to a picture of a molecular orbital. The theorists soon found that several different molecular orbitals could be formed from combinations of atomic orbitals. For instance, you could add wave equations or you could subtract them and various combinations like that. In fact, it is possible to build as many molecular orbitals as atomic orbitals are used in their construction. The molecular orbitals are filled by electrons from the atoms that went into making the molecule, two per orbital, starting with the lowest energy and filling up, though the molecular orbital, like a house on a street, exists whether it is occupied or not. In 1952 in Japan, Kenichi Fukui (1918–1998) found that by calculating the energies of the molecular orbitals and finding the shape of the frontier orbitals, he could predict how one molecule would react with other molecules.[5] As we have mentioned, chemistry happens when electron clouds overlap: When two nuclei with electrons around them get close enough, the electrons of each will feel the tug of the positive nucleus of the other and the push of the alien electron cloud. The colliding nuclei will then either separate or their electron clouds will rearrange to minimize the tensions and form a bond between them. Fukui showed that the shape of the highest energy occupied molecular orbital determines how it will fit, how it will react, with other molecules.

Although Fukui, son of a foreign trade merchant in Japan, reported that chemistry was not his favorite course in high school and his chemical career was basically chosen for him by his father and his father's friends, he no doubt felt new affection for the subject when he received the 1981 Noble Prize in chemistry. Fukui shared the prize with Roald Hoffmann, who developed another important application of molecular orbital theory.

The team of Robert Woodward and Roald Hoffmann found that an interesting class of reactions could be explained in terms of molecular orbital theory. In 1965 they reported that nature's insistence on the conservation of symmetry in molecular orbitals accounted for the behavior of some ring-closing reactions; that is, a reaction where one part of the molecule reacts with another, sort of like someone touching their toes. They found that if the structure of the highest occupied molecular orbital was such that the molecule could react and preserve the symmetry, then the reaction would proceed by merely warming the molecules. If the structure of the highest occupied molecular orbitals was such that the reaction would not preserve the symmetry, then the reaction would not proceed—unless light energy were used to promote an electron to a higher molecular orbital. If this molecular orbital had the right structure, then the reaction would commence. Thus, molecular orbital theory and symmetry elucidated an entire class of reactions that had previously gone unexplained. The situation might be understood by recalling childhood games in which everyone joins hands and forms a ring. As long as the symmetry is right—and as long as everyone has a hand to offer—a solid ring can form. However, if some children are standing on their heads and offer a foot instead, there is no solid ring. Woodward and Hoffmann were saying that the right input of light energy could flip these wayward children and set them to rights, and additionally, by knowing the symmetry of the situation beforehand, one could predict if light energy would be needed.

Robert Burns Woodward (1917–1979) was born in Boston, Massachusetts, and entered MIT when he was sixteen years old, but he was suspended for a term in 1934 for failing to attend classes.[6] After he was allowed back he barreled through and finished his Ph.D. by the time he was twenty. He then moved to Harvard, where he became instructor at twenty-four, assistant professor when he was twenty-seven, associate professor when he was twenty-nine, and full professor by the age of thirty-one. Anyone familiar with academics knows this is not a stellar rise—it is explosive.

Roald Hoffmann's career progressed a bit differently. Hoffmann was born in Poland in 1937. As Jews, he and his family were interned in a labor camp during the war until his father managed to smuggle the mother and child out with the help of a sympathetic Ukrainian. His father attempted to organize an escape for himself and others, but the attempt was discovered and he was put to death along with most of the rest of Hoffmann's family.

After the war Hoffman and his mother left Poland and lived in various displaced person's camps until they managed to make their way to the United States. Hoffmann began college with a medical career in mind but soon concentrated on organic chemistry and then theoretical chemistry. His background in organic chemistry made him the logical choice for Woodward to consult on ring-closing reactions (the touching-your-toes reactions). Together they worked out the correlations between the reactivity and the symmetry of the molecular orbitals. Woodward received the Nobel Prize in 1965 in recognition of his extensive body of work in organic chemistry. Hoffmann would likewise receive a Nobel Prize in chemistry in 1981 and has become a published poet and writer, as well as remaining an active researcher.[7]

In addition to the wonderful results detailed above, molecular orbital theory provided valuable insights into the premier probe of molecular structure: spectroscopy. Spectroscopy, as we've noted, is the technique that gave the first evidence that there is an internal structure to the atom, and now the light would be turned on molecules.

To understand why light interacts with molecules, recall that light can be modeled as an oscillating electromagnetic field. The molecules, with their lumpy peaks and dips of electron clouds always rotating and vibrating (our squiggly jellyfish), should interact with the push and pull of a waving electric and magnetic field. But not all wavelengths of light interact with all molecules. In fact, the exact wavelengths of light that are absorbed by a particular molecule are as characteristic of that molecule as fingerprints. Why are only some wavelengths of light absorbed by the rotating and vibrating molecule but not others? Because molecules aren't jellyfish. They're more like wishbones.

If you throw a jellyfish across the room it will rotate and vibrate as well as fly forward, but a wishbone will not vibrate because the two sides are too stiff to bend and vibrate. Molecules are also vibrationally "stiff." Like the cups in Planck's field, definite amounts of energy have to be put in before the molecule starts vibrating at excited levels. Not just any amount will do. When the full spectrum of light is shown on a molecule, just these particular wavelengths of light will be absorbed. The pattern of discrete wavelengths that a molecule absorbs makes up the fingerprint for the molecule.

Once the quantum nature of molecular vibrations was understood, a long awaited result emerged: the near fit between heat capacity and the kinetic theory of gases. Maxwell and Boltzmann couldn't quite fit the experimentally measured ratio of heat capacities because they knew molecules had to be vibrating as well as rotating and translating, but they had to assume any amount of energy could excite vibrations, just as in classical mechanics. They didn't know about quantum mechanics. Without knowing about wishbones, they had to assume jellyfish. But when the rigidity of molecules was calculated by quantum chemistry, it became apparent that at room temperature there just wasn't enough energy to boost the molecule to an excited vibrational mode. So less energy was being absorbed by the molecule as a whole. When theorists took quantum mechanics and statistical mechanics into account, they got their match with the kinetic theory of gases. The main contributors to the marriage of quantum mechanics and statistical mechanics include Niels Janniksen Bjerrum (1879–1958), one of the few people Jean Perrin tolerated to work with him on his delicate experiments, and John von Neumann (1903–1957), the mathematician who developed game theory, a method by which winnings can be maximized or losses minimized in a competition with an opponent of known means and resources.[8] Neumann's game theory had been used to predict market trends and diplomatic strategies.

Group theory is used to determine the nature of the vibrational motions of molecules: Once the symmetry of the molecule has been established, no vibrational mode can be proposed that would break this symmetry. Symmetry also dictates which light-induced transitions between vibrational states will be allowed and which will not be observed.

Once these vibrational levels are identified, it must be acknowledged that they are quantized, meaning they can only be excited by a certain basic amount of energy (recall the sudden shattering of the wine glass when hit by the right pitch of sound). However, higher vibrational levels can be excited by integral multiples of this basic unit of energy. So, like climbing the rungs on a ladder, the molecule can be excited to higher and higher vibrational levels until it simply flies apart. This idea of a ladder of vibrational energies was used by Hertha Sponer (1895–1968), the co-creator of a classic physical chemistry technique, the Birge–Sponer method, which is used to calculate, from vibrational spectra, the energy required to break a chemical bond. The procedure consists of summing up all the energies required to boost the bonded atoms to successively higher vibrational levels until the bond vibrates apart.[9] Hertha Sponer developed the technique while working with Raymond Thayer Birge (1887–1980), an influential researcher in molecular spectroscopy and quantum physics at the University of California at Berkeley.[10] Sponer joined the department of physics at Duke University in Durham, North Carolina, in 1936 and 1946 she married James Franck (1882–1964), the 1925 Nobel Prize winner in physics and also an important contributor to the theory of molecular spectroscopy. Sponer had come to the United States from Germany about the same time as Emmy Noether. A former physics student at Duke recalls that Sponer was wont to speak excitedly about her work and frequently walked the campus with two Doberman pincers that only understood German commands and were nearly as tall as Sponer herself. This same student remembers being very grateful for having passed the class she took from Sponer, though she finished at the top of her class in the second semester of the course, taught by a different instructor.[11] In the archives of Duke University resides a letter from Robert Millikan (1868–1953), the person credited with first measuring the fundamental charge on an electron, to then president of Duke University, W. P. Few, explaining that he believed Few would "get more for [his] expenditure if in introducing young blood into a department of physics [he] picked one or two of the most outstanding younger men, rather than . . . a woman."[12] President Few evidently discounted Millikan's advice, as Sponer served thirty years on the faculty at Duke.

While group theory helped to sort out vibrational spectra, the electronic spectra of molecules, wherein light promotes an electron from one molecular orbital to another, still posed challenges. In 1929 Gerhard Herzberg (1904–1999) proposed that adding atomic wave equations would form bonding orbitals, whereas subtracting them would form antibonding orbitals.[13] This important concept helped not only clarify the spectra of compounds with electrons in higher energy orbitals (excited states) but also offered an explanation for why oxygen is attracted to a magnet. It also explains why the Starship *Enterprise* had so much trouble with its dilithium crystals.

In bonding orbitals, electrons find themselves spending time between the two nuclei and holding the nuclei together like glue. The electrons are attracted to the nuclei and shield the positively charged nuclei from each other. If an electron is removed from a bonding orbital and promoted to an antibonding orbital then the bond becomes weaker or disappears. Dilithium would have two electrons in a bonding orbital, but if light of the right wavelength hit it, an electron would be promoted to the next closest antibonding orbital and that would be the end of the dilithium bond. Diatomic neon, on the other hand, is never observed because molecular orbit theory predicts that it should have equal numbers of bonding electrons and antibonding electrons. Diatomic oxygen (atmospheric oxygen gas) ends up having two unpaired electrons because there are two molecular orbitals at the same energy, and the two electrons that are available to fill them go into separate orbitals to get as far away from each other as possible. They also have their spins oriented in the same direction. This means there is a net magnetic field generated and that oxygen gas can be trapped by a magnet. Linus Pauling used this magnetic property of oxygen to develop monitors for oxygen levels in submarines in WW II,[14] and this process is still used to monitor oxygen in blood during anesthesia.

Gerhard Herzberg, born in Hamburg, Germany, worked with both Heitler and Lennard-Jones.[15] He originally wanted to be an astronomer but was told he would have to be independently wealthy to pursue that career. So he studied physics and engineering instead and set off on an academic career. In 1935 he had to leave Germany because his wife and collaborator, Luise Oettinger, was Jewish. They were allowed to leave with only ten German marks each, equivalent to $2.50, but they were able to take their spectroscopic equipment and personal belongings.

The Herzbergs fled to the University of Saskatchewan, where Herzberg had accepted an appointment. He pursued research grants (his first was for $1,500), and he and his wife did extensive spectroscopic research, collecting and interpreting the spectra of small molecules of atmospheric importance. His wife died in June 1971. He learned that he had received the Nobel Prize in chemistry later that year.

With the advent of electronic computers in the 1950s the tedious computations required by molecular orbital theory became feasible. Since then molecular orbital theory has become the tool of choice for all applications except one: the simple intuitive explanation of molecular bonding and three-dimensional structure. Here valence bond theory is still very useful and is a well-respected pedagogical device.

So now that the problem of molecular bonding and molecular structure had been developed to a point of a reasonable usefulness, was the job of the physical chemist reduced to tidying up loose ends? Not by a long shot. There remained and still remain major challenges for the physical chemist. Molecules aren't just fluffy white clouds floating freely in the sky. They can also be thunderheads, charged with electricity. This was the problem tackled by the next group we will consider: the Ionists. The Ionists (or, in German, *Ioner*)[16] were named for their radical belief in the independent existence of a charged particle, or *ion*, in solution. Why should this be such a radical stance? Because two oppositely charged particles, like the opposite poles of a magnet, should be attracted to one another. But we never see beakers mysteriously sliding toward each other or slamming together on the laboratory bench. Yet we say we have established the existence of ions in solution. It is because of these current-carrying characters that people are so strongly advised against changing light bulbs while standing in a puddle. So what is really going on and why? Let's look for the answer with the Ionists, and join them as they question why ions don't always slam together but neutral molecules sometimes do: the result of an elusive quality they called affinity.

Die Truckene Trunckenheit.

Chapter 24

The Ionists
and Affinity

> *... theory should be measured by its fruits. Here ... applies the proverb:*
> *"Virtue must show itself."*
> Jacobus van't Hoff to Svante Arrhenius, circa 1900

The Ionists were so named because of their belief in the existence of electrically charged species, that is, *ions*, in solution.[1] But they were much more than that. Just as Isaac Newton found an ingenious and elegant model for his force of gravity, the Ionists sought a similar model for another force—chemical affinity—the force they believed attracted one element to another in a chemical reaction. But even as the Ionists admired Newton's model of gravity, they knew they could not hope for such economy in their model of affinity. Newton, on a nightly basis, could look up into the sky and assure himself of the reality of his objects, but the Ionists had to remain skeptical of the yet unproven atoms. Newton contended with an attractive force that diminished with distance, but the Ionists dealt with forces that were at times attractive and at times repulsive. Newton's planets followed the same force law, winter or summer, but the affinity of the Ionists behaved differently on hot days than on cold. Newton's masterful formulas explained the paths of cannonballs and the orbits of the constant planets, but the Ionists dealt with materials that changed form and identity—sometimes lazily, sometimes explosively.

Given the difficulties, it is easy to understand why a more established, more sober school of chemists saw the Ionists as on a fool's errand. They saw as the mission of the chemists to continue to synthesize and characterize new materials and to leave physics to the physicists. The first inroads into the new physical chemistry, then, had to come from those on the outside looking in. The academics in this new field had to work with lower budgets and carry heavier teaching loads. They had to risk ridicule and even censure. Though photos of these first physical chemists, taken when they were older and more entrenched, convey an aura of authority and austerity, they began as a raucous and rebellious lot. It would be more fitting to see snapshots of them in shirtsleeves, leaning over long tables in dimly lit beer halls, scribbling down equations on scraps of paper, using steins and sausages for models, and arguing far into the night over mass action, electrochemistry, osmosis, and affinity.

In 1864 efforts to put mathematical reins on chemical affinity came from the two Norwegians that we first met in our discussion of the contributions of Willard Gibbs—Cato Guldberg and Peter Waage. They had been friends from youth and became brothers-in-law when they married two sisters. Waage was a chemist and Guldberg was a mathematician, and they both taught for a living. Their goal was to find a mathematical relationship that would describe the course of chemical reactions, and to a fair degree, they succeeded. Newton's gravitational attraction is proportional to mass, so accordingly, they looked for mass to play a role in affinity. What they found was by no means a tidy proportionality between the amounts of the reactants and the force of the reaction. But they did find a constant ratio between the masses of the products and reactants at equilibrium, now known as the *equilibrium constant*. It wasn't exactly what they were looking for, but they liked it. The observation that adding more of one species would cause reaction to occur until equilibrium was reestablished became known as the law of mass action.

An analogy for the law of mass action might be a party in an apartment with three rooms: a living room, a bedroom, and a kitchen. The first wave of people arrives, comes into the living room, but quickly disburses to the other rooms. If the roof were taken off of the apartment, you would see some people in the kitchen, some people in the living room, and some people in the bedroom, and the relative populations would be pretty constant.

Now let's see what happens when a new group of people arrives. This new group again crowds into the living room, but soon distributes themselves into the kitchen and the bedroom. At equilibrium there will be more people in each room, but about the same ratio of kitchen people to living room people and bedroom people to living room people and kitchen people to bedroom people. The party is an example of social mass action. Chemical reactions were said to obey the law of mass action if increasing the concentration of reactant (letting more people into the party) increased the amount of product (people in other rooms) or vice versa. The Ionists were very pleased with their knowledge of this mathematical ratio, but they would have been more pleased to know why it existed. They kept looking for chemical affinity.

Another place the early physical chemists looked to find the force of affinity was in the action of electricity on materials, and for good reason. They knew that positive and negative charges were attracted to each other, so they hoped that electrical charge might account for all affinity. But the current required to produce a certain mass of product didn't seem to depend on the affinity, that is, on how strongly the elements were attracted to each other. For the same amount of current, the same amount of zinc could be collected from zinc chloride as zinc hydroxide, but zinc chloride dissolves readily in water and zinc hydroxide does not. As we now understand the problem, the current is a flow of electrons and the number of electrons required to electrically neutralize positively charged zinc should be the same no matter what it is bonded to. But the early investigators of electrochemistry were loath to accept atoms, let alone electrons.

They also had problems explaining how a current might be carried through an aqueous solution when it wasn't carried at all through air and only very poorly through pure water. Faraday had proposed that when two electrodes are placed in a solution and hooked up to a power source, the electrical strain between them temporarily breaks down the solution into charged particles that can carry current. But in some situations—such as batteries—current can flow even in the absence of an applied potential. This phenomenon is easily demonstrated with a lemon, a galvanized (or zinc-coated) nail, a penny, and a voltmeter. When a penny is stuck in one end of a lemon and a galvanized nail in the other, a voltage is measured across them: a lemon battery. So although electrochemistry offered hope for explaining affinity, there remained many unanswered questions.

Working with solutions many times more dilute than lemon juice, Svante August Arrhenius, in his doctoral dissertation of 1864, made a rather tentative suggestion. He said that current could be carried in solution in the absence of applied potential by certain "active agents" that existed at all times in the solution. Though Arrhenius would not go so far as to name them in his dissertation, these active agents would soon be called ions. Although the empirical evidence was on his side—solutions carry current—he hesitated because he couldn't explain, once separated, why the ions didn't simply come back together again. To ease his conscience on this point, Arrhenius proposed that only a small portion of a material is ionized at any one time. Arrhenius said ionization could be described by the law of mass action.

Svante Arrhenius (1850–1927) was born to a Swedish farming family. When his father obtained a management (nonacademic) position at the University of Uppsala, the family moved to the city. Arrhenius matriculated at Uppsala but after five years there went to Stockholm to work with a physicist on the electrical conductivity of solutions. His dissertation, however, was evaluated at his home institution at Uppsala and his examiners, one chemist and one physicist, passed it, but with such a poor evaluation that he had virtually no hope for a university position. Arrhenius, looking for a second chance, sent his dissertation to Rudolf Clausius, who had once been a young rebel himself, but now the staid professor did not respond. Still hoping, Arrhenius sent his dissertation to a couple of outlanders: Jacobus van't Hoff in the Netherlands and Wilhelm Ostwald in Russia. Van't Hoff read it and responded favorably. Ostwald did better than that; he met with Arrhenius and asked him to join the faculty at his institution in Russia. Arrhenius was not enthusiastic about leaving Sweden for Russia, but with the leverage of an outside offer, he got a position at Uppsala. Then the physicist under whom he had done his conductivity work helped him secure funds to travel to the labs of Ostwald, van't Hoff, Boltzmann, and others. When Arrhenius returned he had more confidence in his results.[2]

Why had Arrhenius's proposal caused him so much difficulty? One reason, as stated, was that the ions, once separated, should recombine, and indeed, a satisfactory answer as to why they didn't would not be found until the early 1920s. But another problem was that nobody knew what to do with the theories. The chemists didn't think of them as chemistry and the physicists didn't think of them as physics. But the physical chemists would give them a home. One physical chemist in particular, van't Hoff, appreciated Arrhenius because Arrhenius helped van't Hoff get a handle on his own weighty problem with affinity: osmosis.

Jacobus Henricus van't Hoff (1852–1911) decided to be a scientist after working one summer in a sugar factory and surmising factory work was not what he wanted to do.[3] In 1877 he was appointed lecturer in theoretical and physical chemistry at the University of Amsterdam, where he began his investigations into osmotic pressure.

Osmosis is the process by which skin puckers in the bathtub and pickles pickle in the vat. When solutions of two different concentrations are separated by a semipermeable membrane (solvent can flow through but solute cannot) the solvent will flow in the direction that will cause the greatest dilution. Skin cells are filled with a concentrated soup of salts, proteins, and other materials. In the bathtub, the bath water flows into the salty solution of the skin cells in an attempt to dilute the cells. Because the skin forms a semipermeable membrane (water can flow in, but the cells cannot flow out) the skin swells and buckles into folds. In pickles, water flows from the cucumber cells into the salty brine in an attempt to dilute the brine and the pickle shrivels. But pickles don't shrivel to dust and our skin doesn't explode in bath water because the solvent flow stops when an opposing pressure is built up within the skin or in the brine. This opposing pressure, the pressure that allows osmosis to go so far and no further, is called the *osmotic pressure*.

As we understand it now, osmosis is an entropy effect—it is an attempt by nature to create more chaos. Dilute solutions are more chaotic than concentrated solutions in the same way that a messy desk is more chaotic than a neat one. The ability to locate something is a measure of order: If you can locate everything on your desk, your desk is ordered; if you cannot, it is disordered. A dilute solution is more disordered than a concentrated solution because it is more difficult to locate the solute in a dilute solution: You can locate and extract a drop of red dye when it is concentrated in a teaspoon, but not when it is diluted in an ocean. Osmosis is important to many life processes and at the time was of interest to physicians, biologists, and botanists; however, van't Hoff was interested in osmosis because he was attempting to get a handle on affinity.

Van't Hoff wasn't thinking in terms of entropy. Van't Hoff's premise was that affinity attracted solvent to the semipermeable membrane or the solute molecules on the other side. What he found, however, was that his dilute solutions seemed to display a gaslike behavior: His osmotic pressure increased as the temperature increased, when kept at a constant volume, just like steam pressure in a pressure cooker. According to the story, the similarity between osmotic pressure and gas pressure occurred to van't Hoff when he ran into a botanist friend on the street outside his laboratory.[4] The friend mentioned he had received a letter from a fellow botanist, William Pfeffer (1845–1920), who was experimenting with the temperature dependence of osmotic pressure.[5] When the botanist relayed that the proportionality between osmotic pressure and temperature was 1/270, a light bulb went on for van't Hoff. He knew that the lowest possible temperature on the gas-law temperature scale, established by Kelvin, was minus 273 Celsius. Could gas laws be related to solutions? He tried using gas-law equations for osmosis data and found a good fit. Osmotic pressure increases with increasing temperature, just as gas pressure increases with temperature. Osmotic pressure increases as more solute is added, just as gas pressure increases when more gas is added. Osmosis had gotten him no closer to chemical affinity, but he had dramatically demonstrated that physics applies to behavior of solutions as well as the behavior of gases. He found a set of straightforward algebraic equations to describe solution behavior.

Van't Hoff soon found that for certain solutes he had to introduce a correction factor, now called the van't Hoff factor, to help them fit the gas-law pattern. What was the origin of this factor? Arrhenius, the struggling postdoctorate he had encouraged, managed to return the favor by explaining the van't Hoff factor. Arrhenius wrote van't Hoff a letter saying that it appeared that the solutes that were causing problems were also solutes that enhanced the conductivity of aqueous solution. Van't Hoff had found that osmotic pressure depended on the amount of solute in solution. If this were interpreted as the number of particles in solution, then this number would be greater if the solute dissociated into ions, which was what Arrhenius had claimed enhanced conductivity. For van't Hoff this was the missing piece to his puzzle and convinced him of the correctness of his theory. Arrhenius wrote up his ideas in the first volume of a new journal, *Zeitschrift für Physikalische Chemie*, which had just been founded by another of Arrhenius's old friends, Wilhelm Ostwald.

Wilhelm Ostwald (1853–1932) was of German ancestry, but born in what was then Russia. At the age of thirty-five he was offered a position at a university in Leipzig, Germany, but only after van't Hoff and three others had turned it down, probably because it carried a heavy teaching load. But Ostwald accepted the teaching load and less-than-lucrative support for his research (he ended up paying for extra assistants out of his own not-too-deep pockets) because Russia was beginning to phase out foreign academics. He was glad for the opportunity for a graceful exit. He also saw it as a chance to pursue research on his ideas on affinity.[6] Ostwald, as we have noted, was intrigued by Arrhenius's ions, especially an extension of Arrhenius's ideas: that conductivity might be a measure of affinity. The basis for Arrhenius's proposition was that solutions of acids and bases make good conductors and the reaction between acids and bases also gives off heat (our lemon battery is an example of an acid solution contained in a lemon skin).

The notion that heat given off by a chemical reaction might be a measure of chemical affinity had been around for sometime. It was assumed that elements attracted to each other should give off energy as they come together because energy is required to keep them apart.

Fig. 27. Wilhelm Ostwald. Edgar Fahs Smith Collection, University of Pennsylvania Library.

The amount of heat given off, it was thought, should be proportional to their mutual affinity. In the mid-1800s Marcellin Berthelot (1827–1907) had stated this idea in the form of a fundamental principle: The only reactions that are spontaneous are the ones that give off heat.[7] However, annoying investigations kept cropping up that found that this was not quite right, and a few dissenters bravely said so. The extent to which this was a downright dangerous stance, however, is demonstrated by the fate of Pierre-Maurice-Marie Duhem (1861–1916). In 1884 Duhem presented a dissertation that followed Willard Gibbs and Herman Helmholtz and argued that the balance of energy and entropy determined the spontaneity of a reaction. In our discussion of statistical mechanics, we illustrated this idea with our theater example: People will tend to choose a theater based on price and seating. They are willing to pay a higher price if they can spread out and be comfortable. Unfortunately Berthelot was still very influential in French science and saw to it that Duhem's dissertation was turned down.

Not only did Duhem have to prepare a new dissertation on a new topic, his career was effectively scuttled.[8] He was never called to a prestigious position in Paris, even though his work warranted it. Van't Hoff, however, already out of the mainstream as it were, was able to develop his own ideas without incurring quite so much hostility. Van't Hoff incorporated heat into the law of mass action and showed how spontaneous reactions could take in heat, too.

This idea can be illustrated by returning to our party for a moment. At equilibrium, there will be a certain number of people in the kitchen, a certain number in the living room, and a certain number in the bedroom. The percentage of people in each room will stay approximately constant, though the total number of people at the party may change. But even though the percentage of people in each room will stay approximately constant, the situation is dynamic, not static. The kitchen people will not stay in the kitchen all night. They will get whatever beverage they want and go into the living room. When they enter the living room, some of the living-room people will feel the room growing a little more crowded and decide to go into the kitchen or bedroom. And so it goes, with people milling from room to room, but, on average, the population of each room staying fairly constant. The situation, as van't Hoff saw it, was similar for chemical reactions. He emphasized the dynamic nature of equilibrium by introducing double-headed arrows into his equations for reactions.

To put this idea into mathematics, van't Hoff adopted an equation found by August Friedrich Horstmann (1842–1929) for the temperature dependence of vapor pressure, and simply said the form also applied to solutions.[9] The beauty of his adaptation was that it showed how the equilibrium concentrations would shift with temperature for both reactions that gave off heat and those that required heat. Van't Hoff's equation gave physical chemists a new level of control over reactions: They could shift the equilibrium toward products or reactants by heating or cooling the reaction mixture. They had leverage. They had numbers. Chemistry was beginning to feel a bit like physics.

But though the new equations meant great progress on one front, the physical chemists still had no direct handle on affinity. Conductivity was not a measure of affinity. Osmosis was not a measure of affinity. Heat was not a measure of affinity. And when it was established that reactions that required heat could be just as spontaneous as ones that gave off heat, some began to question if the concept of affinity was useful at all. So in the end it may be said that the Ionists accomplished their mission, though perhaps not in the manner they had envisioned. The Ionists solved their problem of affinity by forgetting affinity. They did not find one simple law to fill the role of Newton's gravity in their chemical systems, but they found a series of complex, though reliable, relationships that reflected the complexity of the task they had undertaken.

Chapter 25

Intermolecular Forces

We see that wine flows through a strainer as fast as it is poured in; but sluggish oil loiters. This is no doubt either because oil consists of larger atoms or because they are hooked and intertangled . . .

Lucretius, circa 50 B.C.E.

Big whorls have little whorls; Which feed on their velocity; And little whorls have lesser whorls; And so on to viscosity.

Lewis F. Richardson, circa 1950

Of all the complexities faced by physical chemists, intermolecular forces—the stickiness and repulsions of atoms and molecules—has to top the list. People have been thinking about intermolecular forces for a very long time, and understandably so. If it weren't for the matter's stickiness, there would be no condensed phase, and if there were no condensed phase we would not exist. But modeling intermolecular forces is one of the most complicated topics in physical chemistry and has yet to be completely sorted out.

Why are intermolecular forces so complex? To begin with, these intermolecular forces include repulsions as well as attractions. Materials stick together, but they aren't infinitely compressible, either: Trying to compress water in a swimming pool with a belly flop will result in a sore belly. But the dual nature of intermolecular forces is only a minor part of the problem.

Intermolecular forces are electrical in nature, basically the Coulomb interaction we demonstrated with the roll of adhesive tape;

285

the force becomes stronger the closer you get and falls off just as quickly, unlike electrical charges attract and like charges repel. But the situation is much more complex. The kinetic theory of gases states that particles of matter are in constant motion: vibrating, rotating, and flying across the room. Moreover, they are not all flying at the same speed, but a range of speeds. They are not all rotating at the same rotational frequency, but in a range of frequencies. On top of that, materials can form charged species, ions, made from one, two, or many atoms. And in addition to that, molecules are mushy.

One of the most contentious questions for researchers of the 1800s and early 1900s was why water is so sticky. We now know that there are several reasons water molecules stick, one of which is called *dipole* interaction. Molecules made from unlike atoms can form electrical dipoles; that is, they can have a more negative end and a more positive end, so they can behave rather like magnets as far as attractions are concerned. But magnets that are rotating, vibrating, and flying across the room.

Water has a more positive end and a more negative end because it is made of two different kinds of elements, hydrogen and oxygen, and there is an uneven sharing of the electrons in the electron cloud around the nuclei. But liquid nitrogen is made of just one kind of element, and an electron doesn't discriminate between one nitrogen nucleus and another, so there is no reason for uneven sharing of electrons. But liquid nitrogen exists, albeit at a very low temperature (liquid nitrogen is used to treat warts by freezing them). What holds liquid nitrogen together if it isn't monopole or dipole attractions? Quantum chemistry shows that the molecules aren't rigid billiard balls or magnets, they are more like mushy clouds that can deform and distort. These distortions can cause a momentarily uneven distribution of charge, a temporary dipole, even in a spherical atom or molecule, which can be a source of attraction.

So the problem becomes apparent. The molecules are in constant motion. There are attractions and repulsions. There are interactions between ion and ion, dipole and dipole, temporary dipole and temporary dipole—and there are interactions between ion and dipole, ion and temporary dipole, and so on. In mixtures and solutions the situation is even more complex: The magnitude of each of these attractions depends on the species involved. Thus the permutations explode.

But the effort to understand these forces is well spent. The existence of intermolecular forces is responsible for soft towels soaking up water, fluffy clouds forming in the sky, and enzymes carrying out the process of life. Intermolecular forces cause the friction that confounded Leonardo da Vinci and delayed the understanding of motion. Intermolecular forces are extremely important—and their understanding has challenged many the clever mind.

Newton thought there was only one force between particles and that he should be able to model this force with a universal law as powerful as his gravitational law. He knew the force couldn't be gravity: Although he had no real notion of the size of atomic particles, he knew they were smaller than could be seen, even with a microscope, so they were too small for gravity to be a significant source of attraction. Given our current understanding, we agree with Newton. We can calculate that the gravitational attraction for particles of typical atomic mass, at typical interatomic distance, is trillionths of trillionths of trillionths times smaller than the energy of a human heart beat. But Newton also believed the force between particles was different from gravity because he thought the force should be entirely repulsive, which would explain why materials weren't infinitely compressible. He wasn't worried about explaining attractions because he, like others of his day, assumed atoms stood still.

Pierre-Simon Laplace (1749–1827), the mathematician who worked with Lavoisier but kept his head during the French Revolution, was so successful at astronomy (he added corrections to Newton's calculations) that he was emboldened to tackle interparticle forces.[1] He basically accepted Newton's theory of a repulsive force acting at a distance but then added the radiation of particles of caloric and short-range attraction to account for capillary action: the attraction of the narrow pores in a paper towel for molecules of water that allow the towel to pull up the water against the force of gravity. Such was Newton's authority that John Dalton assumed the repulsive force law had been proven beyond doubt. Dalton only added that the repulsions originated from a cloud of caloric surrounding each atom and that the amount of caloric was different for each type of atom. In this way he was able to explain why paper might soak up water without invoking attractions: The water particles were a different size than the paper particles, so the water particles insinuated themselves in the paper particles the way sand fills the cracks in cement. While this theory, like the models of Newton and Laplace, painted a picture that qualitatively reflected reality, there were no quantitative predictions, no numbers that could be measured to confirm or contradict the theory.

Inroads into a quantitative approach were made in the mid-1800s by chemist Thomas Andrews (1813–1885). This Irishman came to chemistry as many did at that time, through the study of medicine and then teaching at a medical school.[2] He noted in his researches (as others had) that Boyle's law—increasing the pressure on a gas will cause the volume to decrease in a regular fashion—failed at high pressures. In fact, at a high enough pressure, many gases condensed. But Andrews went further. He noted that if the gas were heated, then a temperature would eventually be reached above which the gas would not condense, no matter what the pressure. This temperature, which Andrews called the critical temperature, assigned a number to a gas that was characteristic of the intermolecular forces present in the gas.

Initially no one was quite sure what to do with this new number. The theorists working on the kinetic theory of gases, in the main, chose to ignore intermolecular forces. Dealing with what is known as *ideal gases*, that is, gases at high enough temperature and low enough pressure that the intermolecular forces are negligible, they proceeded, though acknowledging the obvious defect in their theory. Ultimately, however, a valid physical model must reflect reality. Hence, intermolecular forces continued to bedevil the kinetic theory of gases. As volume is decreased, the kinetic theory implied, the particles would start running into each other and the gas would be less compressible. But unfortunately the opposite seemed to be true: As pressure increased, the volume for a typical gas would reach a point at which it dropped dramatically and eventually the gas would condense. John Herapath, one of the early authors of the kinetic theory of gases, is said to have been delighted when it was discovered that hydrogen behaved the way it should, even though nothing else had. When Maxwell found that his equations for an ideal gas predicted that the viscosity of a gas should increase with temperature rather than decrease, he was at first skeptical of his equations. He checked them very closely and consulted with friends. It was only after experimentally confirming his result that he seemed to be comfortable with it.

Maxwell attempted a mathematical model for intermolecular interactions that assumed a repulsive force that fell off as the fifth power of the distance. But even as he did this, he knew it was just one of any number of force laws that would work. He hoped it might serve as a computational device even though he had no theoretical basis for choosing it. This technique of finding an equation that fits the data and then trying to explain the equation may seem a bit shaky, but has often proved to be valuable. Max Planck, a few years later, would employ the same method in describing blackbody radiation and would do so with great success. However, Boltzmann rather humorlessly criticized Maxwell for trying such a simple relationship for what had to be a much more complicated problem. Boltzmann knew that there had to be attractions as well as repulsions, though neither man made much progress with the problem.

James Joule and J. J. Thomson threw in their two cents when they observed that a certain class of gases cool when they expand. This effect, the Joule–Thomson effect, is the basis for refrigeration: A working fluid is compressed and then allowed to expand. As it expands, it absorbs heat from the surroundings and uses the energy to break the intermolecular attractions between particles. They found that the temperature of hydrogen was the only one that seemed to increase slightly when it expanded, which can be understood in hindsight. The intermolecular attractions are small for hydrogen, the force of repulsion is much greater, so hydrogen gas is at a lower energy when its particles are spaced farther apart: Heat is released when hydrogen is allowed to expand. In other words, most gases are sticky; they require an input of energy to separate them. Compressed hydrogen is more like a compressed spring. When hydrogen is allowed to expand, it releases energy.

The next advance came from a young Dutch physicist, Johannes Diderik van der Waals. He put forth an equation in his Ph.D. dissertation that Maxwell declared sufficient to win van der Waals's name a lasting place in the annals of science. It did.

The parents of Johannes van der Waals (1837–1923) could not afford to send him to a secondary school that would prepare him to enter a university.[3] At the time universities required a knowledge of classical languages for entry, as well as other basics, which required more schooling than most people received. So on completion of the schooling available to him, van der Waals took a job teaching and studied mathematics and physics on his own. He could not, however, manage to teach himself Latin and Greek, so he still could not meet the entrance requirements for a university. Eventually the government voted to drop the requirement for classical language and at the rather late age of twenty-nine van der Waals entered the University of Leiden. When van der Waals wrote his famous doctoral dissertation, he was thirty-six, had a family, and was teaching secondary school for a living.

The equation that van der Waals proposed, and which won him the praise of Maxwell (it might have remained buried if Maxwell hadn't published an enthusiastic review of it in *Nature*), was compiled from the gas laws of Boyle and the French balloonists. However, he included two correction factors that could be adjusted to fit the behavior of real gases. The breakthrough was that van der Waals provided one universal equation that could be made to fit different gases by assigning each gas its own set of correction factors. These are known as the van der Waals constants for the gas. This pleased the kinetic theorists—there was one equation for all gases—and it pleased the realists—each gas was allowed to behave differently. The equation modeled real behavior: A gas compresses easily until it suddenly condenses, and then it is much more difficult to compress. The beauty of van der Waals's equation is that it modeled the behavior of gases when they were acting ideally at high temperature and low pressure, but also in the region where they weren't behaving ideally—with the same equation. In addition, the correction factors could be related to physical characteristics. One of the factors accounted for the finite size of gas particles, which kept the gas from being infinitely compressible, and the other was a measure of the stickiness of the intermolecular attractions. Van der Waals accomplished two things. He showed that the three states of matter were basically a continuum, not three separate problems, and that there should be a way to liquify all gases: There was no such thing as a "permanent gas."[4]

Van der Waals went on to find a way to measure his constants for various gases using the critical point that Thomas Andrews had described. And he found time to enjoy playing billiards, like the early kinetic gas theorists. But van der Waals knew atoms aren't billiard balls: Atoms are sticky. In 1910 van der Waals received the Nobel Prize for physics.

Though van der Waals had provided the machinery for accounting for intermolecular forces, he did not solve for the origin of the attraction. Another Dutch physicist, Peter Joseph William Debye (1884–1966), made headway in this area.[5]

In 1912 Debye proposed an explanation for the observed fact that some materials were more easily penetrated by an electric field than others; that is, molecules could have a uneven distribution of charge, called a dipole. Debye posited that when molecules had a permanent partial positive charge on one end and a permanent partial negative charge on the other, they could align themselves with an electric field to varying degrees depending on the magnitude of their dipole, which meant they would interact differently in an electric field. Not only did this account for an important intermolecular attraction—the positive end of one dipole could be attracted to the negative end of the other—but Debye also provided a mathematical model that could be used to extract quantitative information from his assumption. By measuring how the material interacted with an electric field, the magnitude of the dipole could be obtained, and from the magnitude of the dipole, the geometry of the molecule might be deduced.[6]

Debye was a prolific worker who won the Nobel Prize in 1936. He trained a generation of physical chemists, though his work was disrupted by World War II. Debye moved to several positions at different universities as was usual for that day. He was at the University of Göttingen at about the same time as Emmy Noether, so they probably crossed paths. He was holding a position in Germany when the National Socialists gained power, but he had not relinquished his Dutch citizenship. Informed by the Nazi government that he could not

Fig. 28. Johannes Diderik van der Waals. Edgar Fahs Smith Collection, University of Pennsylvania Library.

go into the lab unless he became a German citizen, he left Germany and went to the United States. He became a U.S. citizen in 1946.

At first Debye thought all molecules must have some dipole, but as his work progressed, it became apparent that some molecules did not. To account for the existence of attractions even in the absence of a permanent dipole, Fritz London (of Heitler and London valence bond theory) proposed that there may be a temporary dipole that arises spontaneously in the mushy electron cloud of an atom or molecule. Its uneven distribution of charge could induce an uneven distribution of charge in a neighboring molecule, as one magnet can cause another magnet to flip so that opposite poles align. These were the dispersion forces, also sometimes known as London forces. By invoking dispersion forces, even the condensation of noble gases such as helium and neon could be expected and were observed.[7]

However, understanding the origin of the individual intermolecular forces wasn't enough to harness their cumulative effect. In early 1900, however, G. N. Lewis introduced a way of accounting for the net effect of intermolecular attractions in thermodynamic equations. He invented a quantity, *fugacity* (he also came up with the name), that has the same units as pressure, but it is a corrected pressure: corrected for intermolecular attractions. The correction factor, the fugacity coefficient, can be calculated for a given gas from its compressibility, which is also a measure of the intermolecular forces present. *Fugacity* is an old English word meaning readiness to run away or an inclination to flight (the choice of terms provides another glimpse into the poetic soul of the person who came up with the cubic atom). Lewis also came up with the notion of *activity*, a corrected value to replace concentration in thermodynamic equations. Activity has the same units as concentration, but it is a corrected concentration: corrected for intermolecular attractions.

Even though these labors produced some elegant explanations for intermolecular forces, some ways of measuring these forces, and some ways to incorporate them into calculations, there is to this day no way of calculating the net influence of these forces in any one given situation from first principles. In other words, given the structure of the solvent and the solute, it is not possible to plug these parameters into a computer program and have it tell what activity or fugacity to expect. There are just too many variables—vibrational, rotational, and translational motion as well as dispersion, ionic, and dipole attractions and repulsions. Nonetheless, an appreciation of importance of intermolecular forces is absolutely necessary to an understanding of physical chemistry. It was by invoking intermolecular forces that Peter Debye, with his assistant, Eric Hückel (1896–1980), developed a model that finally explained why Arrhenius's ions didn't slam back together again.[8] Debye and Hückel showed that in aqueous solution an ion would be surrounded by a sphere of water molecules, attracted through their dipoles to the charge on the ion. This bulky atmosphere prevented oppositely charged ions from coming back together, which answered the question that had caused so much debate among the Ionists.

The model worked, that is, for fairly dilute solutions. When the concentration of ions got too great, then the situation got sticky again. The Ionists had found their laws by working in dilute solutions, but concentrated solutions are a fact of life. The problem of concentrated solutions was to fall to the heirs of the Ionists.

Chapter 26

Heirs to the Ionists

Any good mathematician can put on the mathematical frills . . . but the result is unsatisfactory if the figure inside is a doll stuffed by human hands, and not a real being of flesh and blood.
Theodore Richards, American physical chemist, circa 1900

The Ionists found an impressive set of equations that model the chemistry of solutions, that is, between chemicals dissolved in liquid. But as one of their heirs would declare, they were dealing with only "slightly polluted waters."[1] A more generous assessment might be that the Ionists were working with idealized solutions, and that they were searching for fundamental relationships that would form the basis for more complicated solution chemistry. But it must be admitted that Wilder Bancroft, the author of the "polluted waters" comment, had a point: There is much more to solution chemistry than dilute solutions. Indeed, nearly all real-world solution systems—blood, rivers, waste streams, manufacturing processes—are concentrated. In their effort to model more realistic situations, the heirs to the Ionists found themselves confronting a murky soup.

Much of the effort to extend the work of the Ionists occurred in the United States. The reasons for the change of venue are legion, but one certain cause was that Europe was already hearing the rumblings of war, which, to some degree, slowed down information exchange. The United States, on the other hand, found itself in a state of political isolation but scientific sharing: The heirs to the Ionists still traveled to Europe to receive their intellectual parenting. When they returned, they brought home physical chemistry.

Many of the generation that would bring physical chemistry to the United States initially went to Europe to study organic chemistry, but were seduced by the excitement of physical chemistry. In the United States there had been a significant lack of appreciation for the ties between physics and chemistry as well as a disregard for mathematics as a preparation for chemistry. Before the end of World War I, it was still possible to get a Ph.D. in chemistry in the Unites States with no more mathematics than algebra.[2] Even at major universities there was either no program in physical chemistry or such a weak program that the labs were located in old water towers and in cellars and professors found themselves financing their own research.[3] But they patiently planted the seeds, and physical chemistry soon began to grow.

Because of the need for independent support, many of the early physical chemists in the United States were a newer version of the gentleman scholar. In 1850, Cooke Josiah Parsons Jr. (1827–1894) dipped into his personal fortune to finance much of the chemicals and equipment that he purchased to outfit his basement lab at Harvard. He was a patient and insistent proponent of laboratory-based education in physical chemistry though he had to work with poor eyesight and hand tremors throughout his life.[4]

Born in Pennsylvania, Theodore William Richards (1868–1928) was the son of a poet and a painter, and his parents tutored him as they traveled through Europe. His advanced education took place at Haverford College and then Harvard. After receiving his Ph.D., he traveled on his own to Europe and studied with leading organic chemists. His chemistry didn't turn physical until he returned to the United States.

Richards's physical chemistry at Harvard consisted of exploring the physical properties of chemicals. Though he carried out other research programs, his most important work was the determination of atomic weights. While this many seem pedestrian, he developed techniques to deal with interferences such as moisture and contamination. It was solid, necessary, elegant work, and it won Richards the Nobel Prize in 1914. He also attempted to study atomic volumes, and in the course of this work came up with his concept of "compressible atoms." However, any development of this theory to a substantial conclusion would have required a mathematical model, and Richards was not strong in mathematics. This same problem of an unsure mathematical grounding was to dog the torchbearer at MIT, too.

At MIT the work was carried forth by Arthur Amos Noyes (1866–1936), who could trace his heritage directly to the Puritans.[5] He retained a bit of the Puritan leanings in that he never married, worked horrendously long hours, and spent his money on research. He originally went on his pilgrimage to Europe with the objective of study in German organic labs, and he eventually found himself in Leipzig admiring Wilhelm Ostwald's new lab. He opted for physical chemistry, which meant he had to teach himself basic calculus and thermodynamics.

When he returned to MIT, Noyes volunteered to pay half of the money needed for a purely research facility if the regents of MIT would pay the rest. They did. Noyes made a significant contribution to the understanding of the physical chemistry of concentrated solutions when he recognized that additional salts can sometimes enhance the solubility of slightly soluble salts rather than depress them, as would be dictated by the law of mass action. Ostwald had made extensive studies on slightly soluble salts, such as sodium bicarbonate (baking soda). As anyone who has used the old standard cure for acid stomach can attest, when sodium bicarbonate is added to water, most of it stays on the bottom of the glass, undissolved. But we know some of it dissolves in the water because the water tastes chalky. Ostwald described the behavior of these slightly soluble salts in terms of mass action: If another source of sodium, such as sodium chloride, were added to a solution of sodium bicarbonate, then the equilibrium would shift in favor of solid sodium bicarbonate. More would come out of solution. For Ostwald's dilute solutions, the description of the equilibrium by mass action worked quite well. But Noyes found that for concentrated salt solutions the situation was, as usual, more complicated. As we now understand it, in very concentrated solutions, extra ions intervene to keep solute particles apart and the salt can be more soluble instead of less soluble. All these complications almost make one yearn for the simplicity of quantum mechanics.

With the success of his research, Noyes became comfortably situated in his Research Laboratory of Physical Chemistry. He answered only to the president of MIT. He ruled the same roost at MIT that Richards ruled at Harvard and Wilder Dwight Bancroft ruled at Cornell.

Wilder Dwight Bancroft (1867–1953) was the author of the "slightly polluted waters" statement given above. Bancroft's family had settled on the North American continent in the 1600s, not far from where the family of Noyes had landed. Bancroft's family was well off to the point that he could decline his salary for several years at Cornell and sometimes he paid assistants out of his personal funds. He played football at Harvard until in his senior year, when he decided he needed a respectable profession, at which point he doubled his efforts until he could graduate with a degree in chemistry. Consequently, his background in mathematics was weak and remained weak throughout his career. On his eventual sojourn to Europe he met and worked for Jacobus van't Hoff. On his return he hoped for a position at Harvard, but the call went to Richards, and Bancroft went to Cornell.

Drawing on the self-assurance implanted in him by his grandfather and his family background, Bancroft confidently began editing the new *Journal of Physical Chemistry*. Not having worried much about building his skill in mathematics, he then set about finding a research program that was more qualitative than quantitative. With this criterion, consciously or unconsciously, he settled on two areas: applications of the phase rule and colloids.

Though the origins of Gibbs's phase rule were mathematical, the relationship was straightforward and algebraic, and the results of its application can be represented on a diagram, a phase diagram. This diagram shows at a glance which phases are present under different conditions of temperature and pressure. The utility of the phase rule was that it allowed one to decide, based on its phase-change behavior, how many components were in a mixture or, given the number of components, determine how the mixture could be treated without having it change phase. An understanding of the value of the phase rule might be derived by looking at a modern application: shampoo. Modern consumers would be upset to find their bottle of shampoo separated into layers after it sat in the shower for a couple of days. What most consumers don't know is that it is a minor miracle that it doesn't. A glance at the ingredients list on the side of a shampoo bottle will give an appreciation for the number of components in a typical shampoo, and they are all in aqueous solution. Anyone who has ever cooked extensively, made Hollandaise sauce or other delicacies, realizes how difficult it is to get organic–aqueous suspensions just right. The shampoo manufacturer has to be able to calculate how many phases will be present (hoping for one) and what conditions will shift the mixture to two phases or more: pressure (can it be shipped by air?), temperature (can it be stored on a loading dock?), or concentration (can the cap be left off in the shower?). A tall order by any account, but one that can be filled if the phase diagram is known or the phase rule can be used.[6]

Phase diagrams have also been put to very good use to predict the composition of solid solutions, such as the alloys that hold up buildings and bridges. These materials age as the metal solutions gradually change. A phase diagram will help predict how ambient conditions will affect the change. Unfortunately, though the phase rule had many uses, Bancroft tended to overextend it. Bancroft's assumption of the position of authority on the phase rule prompted his admirers to salute him as the Phase Ruler.[7] Bancroft was suspicious of the atomic model of matter, as was Ostwald, and he said he liked research on the phase rule because it did not require this assumption. He also used this rationalization when he studied his other main field of interest: colloids.

Colloid means "gluelike," and colloids are aggregates of particles that are much bigger than individual solute particles, but too small to be separated from the solution by just letting it sit or by filtering it. An apt analogy for colloids might be the clumps of people that form in large social gatherings. The test to see if a solution is colloidal is to shine a beam of light through it. If the shape of the light beam is clearly distinguishable, like car headlights in the fog, then the system

Fig. 29. Wilder Dwight Bancroft. Edgar Fahs Smith Collection, University of Pennsylvania Library.

is a colloid. This light scattering by colloids is called the Tyndall effect after Irishman John Tyndall (1820–1893), a contentious and fierce liberal politician as well as natural philosopher.[8] He died when he was sick from insomnia and indigestion and his young wife accidentally administered an overdose of a sleep remedy.

The pioneer of colloid chemistry was Thomas Graham (1805–1869), whom we met earlier in our discussion of the kinetic theory of gas.[9] Graham was the one who found that the rate of gas effusion depends inversely on mass: The heavier the gas, the slower the effusion. He discovered colloids while examining how one liquid diffuses into another. Now we know that homogenized milk and gelatin are colloids. Jean Perrin's gamboge particles, used to confirm the existence of atoms, were colloids.[10] The story is that Bancroft became involved in colloid chemistry after a fire in his lab, and with the small amount of glassware he had left, there wasn't much of anything else he could do.[11] The Tyndall effect can be demonstrated by shining a flashlight through a glass of water to which a quarter teaspoon or less of milk has been added, so he was right about the study not requiring much in the way of glassware.

Unfortunately Bancroft came to some erroneous conclusions concerning colloids, once again trying to overextend the concept. He decided that proteins were colloidal aggregates, not the extremely large molecules that they have been shown to be: The mass of myoglobin, a protein, is almost 1,000 times that of water.[12] In his defense, Bancroft wasn't the only one confused about the subtleties of colloid chemistry. Both Jean Perrin and Albert Einstein theorized that colloids should behave as solutions under dilute conditions (they don't). The problem was that Bancroft's influence, through the *Journal of Physical Chemistry*, made it difficult for researchers trying to put forward opposing views. In addition, he promoted his qualitative chemistry by neatly avoiding articles on the advances in mathematical chemistry and quantum chemistry in the *Journal*. His veto on the review board, it seems, was impenetrable because of his authority and because of his personal control. During the economic downturn after World War I, he kept the *Journal* afloat many times through personal donations.

By the 1930s Bancroft felt he needed to have some conclusion to all his work as nothing substantial had come out of it yet. He believed he had found it when he read a paper by a French physiologist, Claude Bernard, in which Bernard explained that anesthetics worked because of a coagulation of a substance on nerve cells. Bancroft immediately saw colloids. He rushed into print in his journal and others. He and the people he managed to convince described an entire series of nervous disorders including alcoholism, insanity, and drug addiction that could result from colloidal proteins undergoing phase transitions on the nerve cells. He decided that an effective therapy might be to administer a coagulant to cause the wayward proteins to re-coagulate. He announced that sodium thiocyanate, a poison if given in too great a dose, could accomplish the effect and was therefore a near cure-all.

Bancroft subjected his theories to test, even to in vivo trials, with what he claimed to be positive results, and for a time he basked in the limelight. He was written up in *Time* magazine. The American Chemical Society voted him a medal. But others could not reproduce his optimistic results, and the American Medical Association bestirred itself to protest both his methods and his dubious results. As a result, the American Chemical Society hastily asked that he take the medal for his work with the phase rule instead, but Bancroft, insulted all the way around, declined.[13] Bancroft could never manage to back down from his claims. In his defense, it must be said that although Bancroft sometimes advanced by bluster rather than brilliance, he had the courage to tackle very difficult problems.

And so it became increasingly evident in the generation following the heirs of the Ionists that for chemists, a firm grounding in mathematics was critical. G. N. Lewis got a grip on more concentrated solutions by introducing fugacity and activity in a mathematical model. He also tackled the most fundamental challenge for the models of the physical chemist: how to predict the plausibility of a chemical process, in other words, predicting the extent to which chemical A and chemical B might react to make chemical C. Would the equilibrium favor profitable product or lie lazily on the side of unreacted reactants? In grappling with this important problem, Lewis

built substantially on the mathematical theories of a hitherto unac-
knowledged European Ionist: Walther Nernst.

Walther Nernst (1864–1941) was the son of a Prussian civil servant whose earliest ambition was to be a poet, but he became
attracted instead to the romance of physics and chemistry. A colleague of Svante Arrhenius and assistant to Wilhelm Ostwald, Nernst
reached his professional maturity at the time of the blossoming of
physical chemistry in Europe. Walther Nernst occupies a special place
in the history of physical chemistry not only for his contributions but
for the color and flavor that he lends. The staid portraits of a portly
Nernst as a mature scholar belie the hard-living, hard-drinking, outspoken individualist that he was.[14]

When he was in his mid-twenties, Arrhenius reported that Nernst
woke him in the middle of the night dead drunk and that this was not
an isolated incident. In his early thirties Nernst engaged in polemics
with Max Planck and tried to engage Ostwald on his side.[15] Though
Nernst won the Nobel Prize when he was fifty-six, in his forties he was
passed over it because of concerns that the money would be invested
in night cafes in Berlin. But when Emil Rathenau, the German
statesman, industrialist, and tireless worker for the German cause in
World War I, was beaten to death on the streets of Berlin because he
was a Jew, Nernst denounced the death and those who caused it in a
public address as rector of the University of Berlin.

He was equally outspoken and forthright in his science. Of his
substantial contributions to the field, two in particular are of interest
here: his equation to describe the potential of an electrochemical cell
under arbitrary conditions and his enunciation of the third law of
thermodynamics.

The potential of an electrochemical cell depends on the concentration of the reactants; that's why batteries run down. As a voltage-
producing reaction occurs in an electrochemical cell, reactants are
consumed. Nernst not only recognized this fact but found the mathematical relationship that described the behavior. His equation was
essential to Lewis's extensive cataloguing of electrochemical potentials
and accompanying free energies. The equilibrium constants of a host
of important reactions were gleaned from this free energy informa-

tion. But while electrochemistry provided a handle on the behavior of many reactions, there were other reactions, such as combustion reactions, that did not fit well into electrochemical cells. Nernst, however, also helped out in this area, too. It has been pointed out that free energy is the energy free to do work; that is, the heat of the reaction less the amount of energy wasted through entropy. Measuring the heat of many reactions was possible using a device pioneered by Marcellin Berthelot, the bomb calorimeter, which was literally that: a rigid-wall container in which an explosive reaction could take place and the consequent heat measured. Measuring the entropy, however, was not as straightforward. The total entropy could be calculated as the sum of different contributions to the entropy—these contributions might be the chaos introduced by phase change or by changes in pressure or temperature—but to complete the total, the beginning amount of entropy had to be known. This is the same principle used in balancing a checkbook: You can sum the credits and debits, but you don't know how much you have unless you know the beginning balance. In 1905, Nernst, based on experimental evidence, enunciated what is now known at the third law of thermodynamics: The entropy of a pure substance is zero at absolute zero.

The idea is intuitively plausible: At the lowest possible temperature, materials are in a completely ordered state, neatly frozen into position. In reality, there are some difficulties with this picture for certain substances, and the third law had to be revised somewhat, but Nernst had difficulty accepting these refinements. Nernst was a strongly opinionated and assertive individual, but one with a sense of humor. He is said to have stated it had required three people to come up with the first law of thermodynamics, two people to come up with the second, and only one to come up with the third—himself—so there could be no more laws discovered. He was also a sensitive person who was devoted to his family. His wife, Emma, to whom he dictated all his papers, transcribed notes he made on his shirt cuffs before laundering them. He spoke out against the policies of the Nazis when they came to power, which no doubt would have eventually caused him difficulty, but he died in Germany in 1941, while his reputation was still able to protect him.

The newer generation of physical chemists, with their regained respect for the mathematical arts, built substantially on Nernst's third law. A method for calculating the free energies of gas phase reactions followed from the efforts of many dedicated intellects, from both sides of the Atlantic. The elegant and intricate theory involved the symmetry of Emmy Noether, the group theory of Evariste Galois, the quantized energy levels of Max Planck, the statistical mechanics of Ludwig Boltzmann, and the thermodynamics of Willard Gibbs. Physical chemistry evolved into a powerful theoretical system that could predict and explain many aspects of chemical reactivity. But challenges remained.

Chapter 27

Reaction Rates

*As far as the laws of mathematics refer to reality, they are not certain;
and as far as they are certain, they do not refer to reality.*

Albert Einstein, circa 1950

A t room temperature and normal atmospheric pressure, a typical
air molecule experiences about 7 billion collisions per second.
During a chemical reaction, the electron orbitals (that is, the electron
clouds) on individual reactants overlap and coalesce, the way two bubbles will come together and merge into one. The time for the reaction
is determined by the time it takes for the bubbles to find one another.
Once they collide, if they have enough energy, the coalescence is very
fast. During a collision between two chemical reactants, the time it

takes for the electrons in the individual orbitals to readjust themselves into the orbitals around the products (the time for their electron-orbital bubbles to coalesce) is about a quadrillionth of a second.[1] So why aren't all fires explosions and every heartbeat a coronary?

The reasons are easy enough to list. Gas-phase reactants may encounter each other billions of times per second, but solution-phase reactants can be held apart by a sea of solvent and therefore encounter each other far less often. And when the reactants bump, the desired reaction still may not occur if there isn't the right amount of energy. Even if the minimum amount of energy is available, we know from the kinetic theory of gases that not all molecules have the same energy, they have a distribution of energies. So not all encounters will result in a reaction. They also have to come together in the right orientation to be successful, as a key has to be held at the right orientation to fit into its lock. And, if an encounter manages to be successful, it may be only a single step in a series of steps from reactant to final product, any one of which might be a bottleneck.

Many-step reactions can also have mass effects that are counter-intuitive. The early Ionists found that a version of the law of mass action applied to most reaction rates: The higher the concentration of reactant present at the beginning of the reaction, the faster the reaction occurred. This idea makes intuitive sense—the greater the concentration of single people, the higher the rate of marriages. In many cases it is true with chemicals, too: If concentrated vinegar is added to baking soda, the reaction is immediate and vigorous. If the same amount of vinegar is added in a one-to-ten dilution, the reaction is markedly slower. But for some reactions, the inverse can be true. Higher concentrations of some reactants can sometimes impede a reaction. An analogy might be found in an accounting firm: The more paychecks they have to process, the slower they go, not faster.

Reaction rates do not always respond in a predictable way to changes in temperature, either. Granted, the vast majority of reactions speed up as the temperature increases because raising the temperature increases the speed at which the molecules are zooming around, which further increases the number of collisions and the energy of the collisions. However, if a certain step in the reaction produces heat, then heating up the reaction can drive the equilibrium to the product side. In our discussion of the theories of Jacobus van't Hoff, the Ionist, we used the analogy of a party. We said that the people at the party would settle into an equilibrium distribution throughout the rooms. But now if we turn up the heat in one room, they may gravitate to cooler rooms, and the equilibrium distribution will shift. Similarly, for some reactions, heat can drive the reaction back toward the starting materials, away from products, and can slow the reaction.

Still more complications exist. The state of the reactants can influence the rate of the reaction: Liquid gasoline burns; gasoline vapor explodes. Foreign surfaces are always present in a reaction, even if it is just the surface of the container, so they have to be considered, too. Sometimes these surfaces increase the rate of reaction by adsorbing one reactant and holding it in at an advantageous orientation. But the surface can impede the reaction, too, by cooling the reactants on collision or by promoting a reverse reaction.

Given all of the above, it can be seen why the early investigators of reaction rates found themselves wading in deep water. But wade they did. It is important to know when a reaction is going to explode and when it is going to fizzle. And if it does explode or fizzle, it would be nice to know why, so next time things could be different. A lot of credit needs to be given to these early investigators—they sorted out some important clues—but physical chemists are still working on this problem today.

The beginning of a quantitative picture was painted by Ludwig Ferdinand Wilhelmy (1812–1864). He originally studied to be a pharmacist, intending to go into business with his father, but found the pharmacy business dull.[2] At the age of thirty-one he went back to school to study chemistry and physics and received his doctorate at the age of thirty-four. He assumed his first academic position at the age of thirty-seven, but held the position for only five years before taking up the life of a reclusive scholar (having sufficient means from the inheritance from his father's dull pharmacy business). During his brief academic career, he established a mathematical form for expressing the rates of reactions. Rates are basically a measurement of the amount of change in something over a given amount of time: miles per hour, words per minute, kilometers per second. Wilhelmy said that the rate of a chemical reaction is the change in concentration of one of the reactants or products per unit time. (For instance, the change in the amount of alcohol in fermenting wine per day is a measure of the rate of fermentation.) He also said that this change should be proportional to the concentrations of the various reactants in some fashion. For example, the rate at which bread rises could increase proportionally when the amount of sugar is doubled, but increase twice as fast when the amount of yeast is doubled. But Wilhelmy couldn't predict these changes from theory, they had to be found experimentally, and it is worth noting that we still have to find these dependencies experimentally because there are too many variables (intermolecular interactions, orientation dependence, multiple steps) to be able to predict them directly for any arbitrary reaction. Wilhelmy measured the rate of the change of sucrose in an acid solution and found how the rate changes for acid and sucrose, but not much else. His work received little notice until discovered by Wilhelm Ostwald. It did, however, earn him his place in the history of physical chemistry.

When reaction rates were tackled by Ostwald and his cohorts, it was with an eye to getting a handle on affinity, that elusive attraction between molecules that eluded the Ionists. Another chemist–mathematician team—chemist Vernon Harcourt (1834–1919) and his mathematician friend William Esson (1838–1916)—extended the work of Wilhelmy by measuring precisely the mass dependence of the rates of many different reactions, but found the mass dependence was just the tip of the iceberg.[3] In the 1890s Russian chemist Nikolai Alexandrovich Menschutkin (1842–1907) found that the solvent could make as much of a difference as the mass of reactants, but that it wasn't just a viscosity effect.[4] Jacobus van't Hoff, our paragon of physical chemistry, found a correlation between this solvent effect and Peter Debye's dipole moment for the solvent, but he could not interpret his finding.

Van't Hoff's starting premise in his investigation of reaction rates was that affinity was a gravity-like force between the molecules and that the rate should always be proportional to the amount of the principle reagents in the reaction. We could compare van't Hoff's theory to one explaining the rate of deer sightings on a highway: We would assume that the rate of sightings would be proportional to the number of cars and the number of deer. Fewer cars on the road, fewer sightings. Fewer deer, fewer sightings. One would not expect that increasing the number of deer slightly would cause the number of sightings to explode, or that increasing the number of deer would cause the number of sightings to go down, but that was basically what van't Hoff found. While some reactions were well behaved, increasing the concentration of reagents caused some other reaction rates to increase explosively; for other reactions, increasing the concentration of reactants caused the rate to plummet. These observations were correctly interpreted by van't Hoff to indicate that reactions aren't always simple, one-to-one events. There could be some complicated mechanisms at work.

A reaction mechanism is a series of steps that lead from the initial reactants to the final product and can involve any number of twists and turns. For instance, in our accounting firm, the process of cutting checks may begin with the receipt of the payroll, but may involve some back and forth interplay in the form of audits and error checking; it may involve some bottlenecks in the form of slow photocopying machines; and ultimately the rate of work may be limited by factors external to the accounting firm. The post office may also slow things down. So van't Hoff wrote a book on the subject that went through a couple of editions and translations with the help of assistants such as Ernst Julius Cohen (1869–1944), who died in the gas chambers of Auschwitz.[5]

But without information on the size or shape of molecules (and not completely convinced that they even existed) and skeptical of the rate of collisions predicted by the kinetic theory of gases, van't Hoff made little progress in unraveling the actual mechanisms for reactions. However, van't Hoff persevered in his investigations in the hopes of finding some mathematical relationships, which he did. As we noted above, almost all reaction rates tend to increase as the temperature is increased. This is the reason food is refrigerated: The rate of food decomposition reactions increases at higher temperature. Van't Hoff found that for these reactions, the rate constant, as defined by Wilhelmy, increases exponentially as the temperature increases. Though van't Hoff enunciated this mathematical model first, it is often called the Arrhenius equation because Svante Arrhenius worked with it extensively. But van't Hoff did not seem to mind.

Wilhelm Ostwald's primary interest in reaction rates seemed to center on another complication: catalysis. A catalyst is something that changes the rate of a reaction without being changed itself in the course of the reaction. Some surfaces can speed reactions, and it was soon discovered that other substances could sometimes perform that role. Ostwald's contribution was to note that catalysts increased the rate of the reaction, but did not affect the final equilibrium concentrations: At the end of the reaction there was the same ratio of products to reactants, but equilibrium was reached faster. This observation was correctly interpreted by Ostwald to mean that catalysts simultaneously increase the rate of the forward and the reverse reaction. Liquid refreshment might play the role of a catalyst at our party: At equilibrium there will be the same ratio of people in all the rooms for a wet or dry party, but people might tend to move in and mingle more energetically if there is a beverage or two to be enjoyed. A catalyst, likewise, facilitates both the forward and the reverse reactions, so the equilibrium distribution is not affected, but equilibrium is reached faster.

Jean Perrin, and others who had faith in the atomic theory of matter, showed how the rate of collisions between gas-phase molecules could be calculated using the kinetic theory of gases. Perrin, however, as well as others, also proposed what became known as the radiation hypothesis: the idea that certain types of reactions involving only one molecule had to be initiated by absorbing electromagnetic radiation. It was subsequently shown that they were wrong, these reactions could be self-induced by mutual collision, but no real harm was done. In the process of working on the radiation hypothesis, the collision theory was refined and improved.

One person who profited from this theoretical development was Max Bodenstein (1871–1942), the son of a brewery owner, who set up a home lab as a boy and decided on chemistry as a career.[6] For his

Fig. 30. Jacobus Henricus van't Hoff (seated on right) with Frank Wigglesworth Clarke (seated on left) and Harold Bailey Dinn (standing). Edgar Fahs Smith Collection, University of Pennsylvania Library.

doctoral research he was given the assignment of studying the thermal decomposition of gas-phase hydrogen iodide, a simple molecule made up of one hydrogen atom and one iodine atom. Initially he was not thrilled by the assignment, but when he found his results fit very nicely with the kinetic theory of gases, he was quite pleased. He was probably also pleased when his doctorate was awarded summa cum laude. Unfortunately, hydrogen iodide was the last system he would find that would work so simply, and subsequent study has shown that even this one was an accidental, fortunate fit. But Bodenstein was a careful and persistent experimentalist, and his work provided a wealth of valuable information for theorists to ponder.

Bodenstein also investigated the photochemical reaction of hydrogen chloride. Photochemistry, light-induced chemistry, is responsible for sunflowers turning toward the sun and brewers brewing beer in brown glass to keep sunlight from degrading the beer. A few years earlier, Einstein had suggested that light can be treated as a stream of particles called photons. The number of photons that strike a sample can be calculated from the intensity of light and the length of time the light is turned on. The yield of a reaction should depend on the number of photons that are absorbed by the sample. There are many reactions in which this is the case: one photon, one reaction. But for the reaction of hydrogen chloride, this was definitely not the case: One photon started an amazing avalanche that resulted in 10,000 or more reactions.

What was going on? A chain reaction. The first photon created a radical: a highly reactive atom or molecule with one unpaired electron. Most people are familiar with the term *chain reaction* and use it to describe everything from nuclear reactions to accidents on highways. The highway analogy is a good one: The effect of the reaction multiplies as one reactant creates two reactants, etc., just as one car will cause another car to hit two cars, etc.

In the 1920s and the 1930s, theorists began applying the considerable power of quantum mechanics, thermodynamics, and statistical mechanics to the problem of reaction rates, in particular Mihály Polányi (anglicized to Michael Polanyi) and Henry Eyring.

Hungarian Michael Polanyi (1891–1976) had a businessman for

a father and a left-wing Marxist for a mother. Polanyi originally studied medicine, but even at the age of nineteen showed an inclination to the more experimental side of the art. In the same year as he received his medical license, he submitted his doctoral dissertation on the adhesion of gases to surfaces, and three years later he married a fellow scientist, a chemist. He served as a medical officer in the Hungarian army, then accepted a position at a university teaching physical chemistry. He eventually joined the Kaiser Wilhelm Institute for Physical Chemistry.[7] In 1931, working with U.S. doctoral student Henry Eyring (1901–1981), he made an attempt to exactly describe a single reaction from first principles. They constructed a three-dimensional plot of the energy of interaction that would be felt by a hydrogen atom approaching a hydrogen molecule in the course of a reaction. Although the system is the simplest possible atom–molecule reaction pair, the calculation was important because it was the first of what is now called *potential energy surfaces* to be solved. Later Polanyi developed a reaction-rate theory based on the fleeting, delicate complex formed at the instant of collision called the *transition state*. Back in the United States by then, Eyring developed a very similar theory.

Henry Eyring's father was a cattle rancher in Mexico whose own family had emigrated from Germany in the 1800s. Eyring studied mining engineering for his Bachelor's degree and received a Master's in metallurgy, but after a short stint as a metallurgist in a copper smelter, he decided to go back to school and study chemistry.[8] He joined G. N. Lewis's group at Berkeley, by now a virtual breeding ground of excellent chemists, and received his Ph.D. in 1927. In 1934 Eyring submitted his paper on his version of the transition state theory to the *Journal of Chemical Physics*. The editor at that time rejected it on the basis of one reviewer's evaluation. But an English physical chemist who had worked with Max Bodenstein spoke up on behalf of Eyring's paper and it was published.

Eyring presented a method for calculating a rate constant from first principles. Eyring assumed that the transition state vibrated just like stable molecules, but that only one of its vibrational states led to a reaction. This vibrational state, like all good molecular vibrations, is quantized, so Planck's constant came into the equation. The probability for this state could be found statistically, so Boltzmann's constant was also invoked. An equilibrium would exist between the reactants and the transition state product, so there was also a thermodynamic equilibrium constant involved. To account for the fact that even those that hit the right vibration would not necessarily progress to product, there was a "transmission coefficient" factor; that is, he introduced a constant that can be adjusted to account for more exotic reaction behaviors, should they be present. Thus, in some ways Eyring's equation can be seen to summarize physical chemistry: It involves thermodynamics, quantum mechanics, statistical mechanics, a rate constant, and, in the transmission coefficient, a little wobble room to allow the mathematics to accommodate reality. The value of the equation is not necessarily in the number it predicts, but in its form. By fitting measured rates to the theoretical form, conclusions can be drawn about the shape and nature of the collision complex.

Another investigation into the elusive transition state stemmed from the work of Jacob Bigeleisen and Marie Mayer. Mayer coauthored their paper on the effect of isotopic mass difference on reaction rates, then moved on to other projects, as we shall see when we meet with her again. Jacob Bigeleisen (b. 1919), however, made the study his life work. Bigeleisen turned to the study of chemistry in the Depression of 1935 in hopes of becoming a dye chemist in the local textile plant.[9] But the war effort turned him to physical chemistry and investigations of transition states.

Mayer and Bigeleisen were attempting to find ways of separating uranium isotopes. Separation is a very necessary step in weapons-material research because only one isotope of uranium is fissionable and it must be in a concentrated form. The premise of their work was that the compounds of the two important isotopes of uranium should have different vibrational frequencies because of the difference in mass. Heavy, wet cloth moves less in the wind than light, dry cloth. They used quantum mechanics to calculate the difference in vibrational frequency based on mass difference and the methods of statistical mechanics to calculate energy, entropy, free energy, and finally equilibrium constants for reactions that they hoped would concentrate one of the isotopes in a particular compound. In their work they had to consider not only mass differences in isotopes but also symmetry differences that might be introduced by the isotopic substitution (the sort of difference that might be caused by having one leg slightly heavier than the other).

Bigeleisen's method of using isotopes to probe transition states was necessarily indirect. In gas-phase reactions, the transition states, those fleeting states that are not reactants but not quite products yet either, can be so short-lived that only recently has a direct experiment probe been found. In 1999 Egyptian-born Ahmed Zewail performed Nobel Prize–winning work in which he used light pulses a few quadrillionths of a second long to probe transition states. A first pulse photochemically initiated the reaction and a second served as a spectroscopic probe of the transition state. But Zewail, though making much headway toward the understanding of chemical reactions, was working in the gas phase and with a rarefied gas at that, which is a long way from the dirty world in which we live. However, progress has been made toward understanding reaction rates in solutions and solids, too. In fact, impressive progress has been made toward understanding reactions in human physiology—and there can be no soupier soup than living organisms.

Leonor Michaelis (1875–1949), from a German Jewish family of modest means, was a medical student who found himself better in the lab than at surgery and gynecology.[10] His skills won him an assistantship with Paul Ehrlich, the famous bacteriologist who found the first chemotherapy since mercury to treat syphilis. To fulfill his obligation, Michaelis studied dyes, physical chemistry, and mathematics. He went into practice for a while and then to the University of Berlin, but he did not attain a permanent position or laboratory space. However, he established a small private laboratory with a chemist friend where they proceeded to do research. The quality of the research soon won them sufficient recognition that they attracted students. Working with Canadian guest Maud Leonora Menten, he published a paper in 1913 that provided a now-classic Michaelis–Menten mathematical model for rates of enzyme catalyzed reactions. Michaelis was eventually invited to teach at the University of Nagoya, Japan, where he was given laboratory facilities and worked with research students.

He subsequently moved to the Rockefeller Institute for Medical Research in New York and went on to do important work in the physical chemistry of dyes and proteins, as well as study the new quantum mechanics and lecture on it. Maud Menten went on to be one of the first Canadian women to receive a medical degree and eventually became chief pathologist at Children's Hospital in Pittsburgh. She always maintained a strong research component to her work, integrating physical chemistry methods and models into her study of biochemical problems. She was a pioneer in the use of electrophoresis, a technique that separates molecules on the basis of their rate of migration through a gel under the influence of an applied electric field—the technique later used by Linus Pauling and coworkers to isolate the cause of sickle cell disease.

While it may seem surprising to learn that Pauling, a physical chemist, did research on biological materials, in the next section we will see that one of the most important applications for physical chemistry has been always been physiology. The human body is a physical chemical machine.

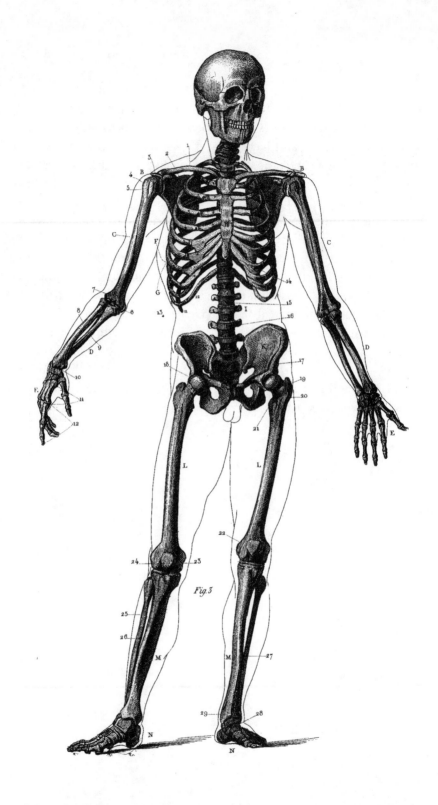

Fig.3

Part VI
The Fruits of Their Labors

INTRODUCTION

Physical chemistry is the chemistry of the future.

Wilhelm Ostwald, 1887

A physical chemist does what a physical chemist does.

G. N. Lewis, circa 1930

By the mid-1900s, physical chemistry was firmly entrenched in college chemistry curriculum. Nowadays the electron-dot structures of Gilbert N. Lewis, the osmosis of Jacobus van't Hoff, the valence bond theory of Linus Pauling, and the free energy of Willard Gibbs can be introduced in high school chemistry. The mathematical models of chemistry, decried in 1830 by Auguste Comte as being "profoundly irrational and contrary to the spirit of chemistry"[1] have become so

319

important that algebra and precalculus are prerequisites for college chemistry, and calculus is necessary for a chemistry degree.

Chemistry is part and parcel of medicine, geology, astrophysics, biology, agriculture, structural engineering, and more, and physical chemists now find themselves involved in all of these fields. Unfortunately, space prohibits our detailing all the areas to which physical chemists contribute. Therefore, in these closing chapters, we will content ourselves with a look at the applications of physical chemistry to physiology, nonlinear dynamics, and nanotechnology and take a glimpse at the challenge of unsolved issues in quantum mechanics. Many of the advances in recent years have been made possible because of new technologies such as computers and lasers, and an extension of the study of the interaction of light with matter, spectroscopy.

Just as photography enabled the scientists of the 1800s to record exactly what their eyes saw, spectroscopy excites scientists today because it allows them to see what their eyes cannot. As may be recalled from our earlier discussion of spectroscopy, it has long been known that light causes changes in matter and matter causes changes in light. In the mid-1800s it was recognized that the wavelength of light that causes certain changes can be very specific: A photograph cannot be developed in daylight because sunlight exposes all of the film, but a photographer can use a red light to see in a "dark room" because red light does not expose the film. When it was realized that a listing of the wavelengths of light that interacted with a specific material, or spectrum, could be used as a fingerprint of that material, spectroscopy became a powerful tool for chemical identification.

But beyond identification, when Max Planck, Albert Einstein, and many others had put the phenomenon on a theoretical footing, it was possible to derive information about molecules from the wavelength of light with which they interacted. For instance, molecular vibrations can be modeled as two balls on a spring. By knowing the mass of the atoms that make up the molecule and the energy of the light that causes them to vibrate, the strength of the spring—the chemical bond—that connects them can be calculated. The rotation of a molecule can be modeled as a Frisbee or discus, and by knowing the energy of light needed to excite a rotation, the diameter of the discus—the bond length—can be calculated. It turns out that ultraviolet radiation can stimulate the movement of electrons from orbital to orbital (the mechanism by which the sun's ultraviolet light can burn skin); microwave radiation can stimulate rotational motion in molecules (the mechanism by which microwave radiation cooks food); and infrared light is of the right energy to stimulate vibrations (the mechanism by which you feel a warm stove burner before touching it). But not all vibrational modes of all molecules can be excited by the absorption of infrared light. If there is an even distribution of electrons in the molecule—if there are none of Peter Debye's permanent dipoles or dipoles caused by vibrations—then there is no molecular field with which the oscillating electromagnetic field of infrared light can interact. But the vibrational modes of molecules with no dipole can sometimes still be detected by looking at how they scatter light.

Light scattering is a phenomenon that affects us on a daily basis: The scattering of light is why the sky is blue. The shorter the wavelength of light, the more efficiently it is scattered from the sunshine, so the high-energy, short-wavelength, blue light is very efficiently scattered. In fact, it is scattered all over the sky, making the sky a baby blue. Some vibrational energy can be subtracted or added to the light in the course of scattering, which can make scattering a probe of vibrational motion. This scattering was first experimentally observed by Chandrasekhara Venkata Raman (1888–1970) of India, and for this discovery he became Asia's first Nobel Prize recipient.[2]

It is a tribute to the patience and perseverance of Raman that he was able to see the effect, because it is very small. However, scattering became a viable spectroscopic technique with the advent of the laser. A laser can provide a very intense light, so even though the percentage of light scattered is small, enough is scattered to be detected. The laser also helped make possible another important new spectroscopic technique: Fourier transform spectroscopy.

Fourier transform spectroscopy is based on the mathematics of Jean Baptiste Joseph Fourier (1768–1830), a near-victim of the French Revolution. However, Robespierre proceeded him to the guillotine, the political tide turned, and Fourier walked away intact. Fourier described how one might mathematically sort out the relative intensity of a particular wavelength signal from a background of many wavelengths, as the ear may pick out the sound of the violins in a symphony or the sound of a child's voice in a crowd. The spectroscopy based on the mathematics, however, had to wait until the advent of lasers. The trick involves sliding a mirror back and forth very rapidly to change the pathlength of the light, and the laser tracks the position of the mirror. The advantage to the Fourier technique is that it allows all wavelengths of light to be shown on the sample at once and the change in specific wavelengths to be sorted out later, which amounts to a tremendous savings in time and in the equipment needed to break the light up into its component wavelengths with a prism or other wavelength selection device. However, the mathematics to sort out the changes in individual wavelengths was so time-consuming initially that any time saved in the experiments was lost in the computation until we had computers.

In addition to making the mathematics of the Fourier transform more tractable, computers also made possible another important advance in physical chemistry: molecular modeling. Molecular modeling, also called computational chemistry, computer-aided chemistry, or computer-aided molecular design, is a method for finding the three-dimensional shape of molecules and for using knowledge of their shape to predict reactivity. The idea behind molecular modeling is that certain characteristics are similar for the same group located on different molecules. For instance, a length of a bond between two carbons can be considered to be approximately the same from molecule to molecule. The van der Waals's repulsions between nitrogen and oxygen on an amino acid will be approximately the same from protein to protein. The results have been remarkable.[3]

Nowadays such calculations can be carried out on a desktop computer or even a laptop for smaller molecules. More complex calculations have enabled researchers to design drugs to enhance or interfere with the activity of biologically important molecules.[4] Though the method had been suggested in the 1930s, the actual practice had to wait for computers to make the required calculations practical. But as we paint these pretty pictures of molecules on computer screens, we again come up against the question that plagued Jean Perrin. Are the calculations reflecting something real? Or are they just numbers? Wouldn't it be better if we could actually "see" molecules? In the twenty-first century, we believe we can.

The smallest object the unaided human eye can distinguish is a tenth of a millimeter. With a light microscope, the microscope used in biology classrooms, one can see objects one hundred times smaller by shining light through a thin section of material or by reflecting light off the surface. In 1933, twenty-six-year-old doctoral candidate Ernst Ruska (1906–1988) broke that barrier by using electrons rather than light to image objects and achieved a resolution of one-millionth of a meter.[5] At first the new instrument, the transmission electron microscope (TEM), did not receive much notice because its beam burned the material it was imaging. But improvements allowed Ruska, his brother, and another colleague to take pictures that provided evidence for the existence of viruses. (But knowing they are there does not tell us what color viruses are: As soon as we abandoned light, we abandoned color. All of the images taken by nonoptical methods, such as TEM, lack information on color. The pictures in magazines and journals are often color enhanced to bring out detail, but the color is purely at the discretion of the artist.)[6]

The TEM method has its limits. For one thing, the electrons have to travel in a vacuum and the sections that they travel through must be very thin (about a thousandth of the thickness of a page of a book) because electrons are not able to penetrate to the same extent as light. Another method uses reflected electrons rather than transmitted electrons, the scanning electron microscope (SEM), which was introduced in 1966. This method allows thicker samples to be imaged, but it does not have the resolving power of the transmission electron microscope. In the 1980s, however, the scanning tunneling microscope (STM) was invented, which allows imaging of down to a tenth of a billionth of a meter (a unit known as an Ångstrom, named for the nineteenth-century Swedish spectroscopist who determined methods for measuring the wavelength of light on the thousands of Ångstrom scale). This new advance in imaging took an entirely different tack. This new imaging technology relies on the ability of electrons to find themselves where they're not supposed to be: a phenomenon called *tunneling*.

Tunneling refers to the ability of an electron to penetrate a barrier that classical physics, cannonball physics, says it shouldn't. In other words, no matter how many times handballs have been slammed against the walls of handball courts over the several millennia that variations of the game have been played, not once has a handball escaped the court by penetrating the wall. But if the handballs were electrons bouncing around in a metal needle tip, once in a while an electron would pop right through the wall of the metal-tip court. The explanation for this electron tunneling is that the equations that describe the motion of electrons can be interpreted as probability waves. Though the probabilities may be very, very small for an electron to penetrate a barrier, they are not necessarily zero, and occasionally an electron will tunnel right through. For electrons leaving the tip of a fine stylus being moved over a surface, the probability for tunneling from the stylus depends on the distance of the stylus from the surface. So by scanning the tip over a surface, the hills and valleys of the surface can be detected, and in fact, the hills and valleys of atoms have been detected. Heinrich Rohrer and Gerd Binnig inaugurated the technique of STM in 1981, and in 1986 Russell Becher and his coworkers were able to move atoms around on the surface using the electric fields of the tunneling microscope.[7] In the famous spelling out of the IBM logo with xenon atoms cooled to within a few degrees of absolute zero, Ernst Mach's challenge, "Have you seen one?" was answered. Yes.

Chapter 28

Physical Chemistry and Physiology

NOT SO STRANGE BEDFELLOWS

Physical chemistry is power, it is exactness, it is life.

Sinclair Lewis, circa 1925

The opening quote for this chapter doesn't come from a physical chemist, or even a chemist. These are the words Sinclair Lewis, in the novel *Arrowsmith*,[1] put into the mouth of Max Gottlieb, professor of bacteriology at a pre–World War I medical school. The words are spoken to Martin Arrowsmith, aspiring first-year medical student and protagonist of the novel. Gottlieb goes on to explain, "If you are not going to be a cookbook bacteriologist, like most of them, you must be able to handle some of the fundamentals of science. All living things are physico-chemical machines. . . . How can you make progress if you do not know physical chemistry? . . ."[2]

For anyone who has leafed through a modern textbook of medicine, this statement may seem a bit odd. The pages are replete with equations and graphs long familiar to the physical chemist. But this has not always been the case. A textbook of medicine from the early 1900s reveals a few graphs of the rhythmic variations of blood pressure or a graph of the change in body temperature over the course of a disease, but little of the numbers and formulas that now seem so necessary for diagnosis and treatment. The transformation of medicine required young rebels, as portrayed in Martin Arrowsmith, to take the lead and battle disease in the laboratory patiently with the "mysteries of freezing-point . . . osmotic pressure . . . [and] mathematical laws which strangely predicted natural phenomena."[3]

Martin Arrowsmith was a fictional character, but in the history of physical chemistry there are ample examples of real-life analogues. Physiology is the science that attempts to explain physical and chemical processes in the body. At one point, the prevailing sentiment was that the whole organism could never be entirely understood by looking at its parts. Physiologists spoke in terms of vital forces[4] and special biological processes that were somehow beyond the normal descriptions of chemistry and physics. However, in the mid- to late-1800s there grew to be another school of physiologists who saw connections between physical chemistry and physiology: the thermodynamics of body heat, the structures of biomolecules, the rates of enzymatic reactions, the diffusion of gases and water through cell walls, and the law of mass action in immunity reactions. Called "followers of Helmholtz," they believed in the application of quantitative science to medicine; that is, they utilized the principles of physics and models of mathematics to describe chemical processes in the body. Some experimental physiologists, the mechanical physiologists as they were sometimes called, believed all biological processes, including thought and will, would ultimately be reducible to chemical and physical actions. In fact, the real-life character that Sinclair Lewis used as his model for what he believed a physiologist should be was Jacques Loeb (1859–1924). Loeb started out as a philosopher, looking for the origins of will, but decided he was more apt to find it in physics than philosophy.[5]

The son of a successful businessman, Loeb switched his studies from philosophy to science when he found the philosophers at the University of Berlin to be "wordmongers." He completed his medical degree in 1884, with a thesis on the relationship between specific brain injury and blindness, though he also felt this work was still removed from what he was attempting to understand. However, when he found some readings on plant tropism, the mechanical response of plants to external stimuli such as gravity and light (the mechanism by which roots know to grow downward and flowers know to open to the sun), he realized he had found his area of interest. He would show tropism in animals, thereby demonstrating a basis for what people had, in his estimation, been mistakenly calling "will," but were actually responses beyond conscious control.

Loeb met with impressive initial success. A breed of caterpillar that climbs to the top of branches had been previously thought to be driven to climb by the instinct for self-preservation. Loeb was able to show that they were merely climbing toward the sunlight. When he placed a light source at the bottom of the plant they climbed downward and stayed there until they starved to death, ignoring the buds on the dark top of the plant.

In another revolutionary study, he showed that it was possible to stimulate unfertilized sea urchin eggs to develop to the larval stage by manipulating the salt content of the solution in which they were immersed, thereby reducing the activation of an egg to a physicochemical phenomenon.[6] However, these early followers of Herman Helmholtz ran into resistance just as the Ionists had: Some physicians saw medicine as more of an art that a science and were loath to apply the rigorous techniques of experiment and control.

In the United States, the idea of an intimate relationship between physical chemistry and physiology found fertile ground. Rudolf Otto Anselm Höber (1873–1953), who was forced to emigrate from Germany by the anti-Jewish laws, used van't Hoff's concept of osmotic pressure and Arrhenius's concept of conductivity to describe how materials move in and out of cells. After making a study of Walther Nernst's textbook of physical chemistry and incorporating physical chemical principles into his interpretation of physiological processes, he wrote his own textbook, *Physical Chemistry of the Cell and Tissues*.[7]

During the time of the first World War and between wars, many of those trained in physical chemistry entered the field of physiology. In the 1930s, Edwin Cohn (1892–1953) headed the department of physical chemistry at Harvard Medical School. He had established a protein research program based on the principles of physical chemistry, but he was initially noted more for his pedestrian approach than any particular discovery.[8] But the patience and care that were the source of derision became his special gift: When the war demanded, he was able to develop a method to separate serum albumin using alcohol. Serum albumin is the protein part of serum that maintains the osmotic pressure of the blood and is used as a substitute for plasma in the treatment of shock. Cohn's contribution has saved an untold number of lives over the years of its use.

In Europe, the fortunes of physical chemistry and physiology were linked in another way: the determination of the structure of complex biological molecules. Wilhelm Röntgen's discovery of X-rays had found application in the determination of anatomical structure, and now it would find its use in atomic-level determinations of structure.

In 1912, Max Laue (1879–1960), a German physicist, considered the phenomenon of light diffraction and realized that if he could find a diffraction grating fine enough, he should be able to diffract X-rays, too.[9] When he realized that a crystal would make a fine enough grating, he proceeded to produce a diffraction pattern by allowing X-rays to impinge on a crystal of salt. The next year, 1913, William Henry Bragg (1862–1942) and his son William Lawrence Bragg (1890–1971) demonstrated that measurements of the spacings of lines in the diffraction pattern could be used to calculate the spacings between atoms in a single crystal.[10]

The father sprang from a humble but honest English family. Bragg did well in school, but was not very productive in his subsequent academic career at the University of Adelaide in Australia. Then in 1903, at the age of forty-one, he became intrigued with the exact nature of X-rays, which still had not been decisively determined. It had been supposed that X-rays were particles or waves or transverse pulses in the hypothetical ether. His rapid series of experiments and publications resulted in a position in England and he returned in 1909. There he and his son worked together to sort out the significance of the patterns of light generated in X-ray diffraction.

William Henry Bragg's son, William Lawrence Bragg, started out working for J. J. Thomson on a different problem. However, he soon became intrigued by his father's line of inquiry. With his father, he worked out the equations that could be used to calculate the distance between atoms in crystals from X-ray data. The publication of the formula appeared in 1913, before the son's twenty-third birthday. Father and son won the Nobel Prize in 1915 (a first), and the son is the youngest recipient to date. The son held the title of Nobel laureate for fifty-six years (another record).

The technique they developed for analyzing crystals with X-rays is known as X-ray crystallography and may be understood by reconsidering the laser-pointer–ruler demonstration that we described in our discussion of electromagnetic radiation. In that demonstration, a diffraction pattern, a pattern of light and dark bands, is generated by the light of the laser pointer reflected from the ridges of a metal ruler. The same thing happened with the Braggs' crystals. The difference is that the crystals are an atomic scale "ruler" and very short wavelength X-rays are used instead of visible light. The Braggs got their good results from sodium chloride, which is a fairly simple cubic crystal. But not all crystals are so simple. A body of mathematical theory had to be derived, using the tools of group theory and symmetry, to provide a general method of analysis. Kathleen Yardley Lonsdale (1903–1971) came to the attention of the elder Bragg through her competent performance as a student of mathematics and physics, so he invited her to work with him in his laboratory.[11] Working with William Astbury (1898–1961), a chemist, she developed a theory of X-ray diffraction that could be used for general structural determinations.[12] In 1945 she was the first woman elected a Fellow of the Royal Society, though she also spent a month in prison for her pacifist convictions during World War II. William Astbury, who worked with Lonsdale on the development of a theory of X-ray diffraction, used the technique in 1930 to examine a single strand of wool, thus opening the door for the use of the technique for biological molecules. His premier finding was that he obtained two different patterns depending on whether the wool was stretched or relaxed, which implied that the molecules of the wool were somehow folded in on themselves. From this initial study began the search for the three-dimensional structure of biological molecules.

John Desmond Bernal (1901–1971) was the son of an Irish squireen, which is a position intermediate between farmer and squire, and his father was at least well-off enough to send his son to Cambridge.[13] Afterward Bernal worked for the Braggs at the Royal Society and then secured a position back at Cambridge, being put in charge of their new X-ray diffraction facility. The account has it that Bernal was a gregarious redhead who went by the nickname Sage and who embraced communist politics and a free-wheeling lifestyle. He once defused a bomb in Liverpool Street Station though the place was evacuated and the bomb squad was on the way. Though he married and remained married to the same woman his whole life, he was notorious for his romantic interludes.[14] Bernal's two great achievements in science were to solve the structure of graphite (it forms flat sheets that are free to slide past one another, which is why it is a good lubricant) and to recruit Dorothy Crowfoot Hodgkin to X-ray diffraction.

Dorothy Hodgkin (1910–1994) was born Dorothy Mary Crowfoot in Cairo, where her family was located while her father, an archaeologist, worked for the Egyptian Education Service. Her father's work was a family affair; her mother became an authority on ancient weaving and Hodgkin herself almost gave up chemistry for archaeology, though her first love won out. She found chemistry through a home lab and was allowed to take chemistry with the boys at school. She became an experimentalist, working with delicate and intricate instrumentation, though at an early age she had been afflicted with rheumatoid arthritis. Consequently her hands were severely malformed and painful most of her life. She studied chemistry at Oxford and was encouraged to follow her inclination to learn X-ray crystallography.[15]

From Oxford she went to Cambridge to work for Bernal. Together they discovered the technique that would make X-rays of large, complicated biomolecules possible: They left them in the solution from which they had been crystallized to ensure they were well structured enough to give a sharp pattern. Biological molecules are meant to hold their shape in liquid, namely body fluids. Once dry, they deform in unpredictable ways and break up the periodic structure that is necessary for X-ray patterns.

As Hodgkin's biographer, Georgiana Ferry, relates, the relationship between Bernal and Hodgkin was intimate as well as intellectual.[16] This union no doubt gathered strength from their mutual passion for science, unrestrained embrace of life, and refusal to be tethered by convention. Though the physical aspect of the relationship ceased when Hodgkin opted for a settled life, they remained close colleagues and friends.

After her work with Bernal, Hodgkin returned to Oxford to build up an X-ray crystallography lab there. During this same period she met and married her husband, a historian. Her labs at Oxford consisted of a hodgepodge of rooms and a hodgepodge of workers including students, assistants, and visitors. But all help was welcome. In her time, the calculations to determine the structure of molecules as complicated as the ones she worked on, such as penicillin and vitamin B_{12}, required months of slide rule and pencil time. Nowadays these same calculations are handled by computers in minutes, but the samples are still prepared by talented human hands. In 1964 Dorothy Hodgkin became the only British woman to receive a Nobel Prize, but it is easy to see why she was selected for this honor: If anyone could pull it off, Hodgkin could. She purportedly sent an assistant to meetings of the organic chemists so that the assistant could interject into the conversation how the structure of vitamin B_{12} had been found—and in whose lab—so that her work would not be appropriated. Once, when asked to leave a scientific meeting that was held in males-only quarters, she refused and had to be carried from the room. She even took on Margaret Roberts, who later became Margaret Thatcher, Prime Minister of Britain, as a student.[17] Dorothy Hodgkin also worked out the structure for insulin, though the effort took many years from start to finish.

The story of insulin makes an interesting sidebar to our narrative on the connection between physical chemistry and physiology. Prior to the mid-1920s, a diagnosis of diabetes was a death sentence, a slow, painful death from starvation. The cause of diabetes is an insufficient supply of the hormone insulin, which is needed to regulate the metabolism of sugar. In 1922 the idea for a treatment came from a Canadian country doctor.[18] At the time it was fairly well known that there is a substance produced by the pancreas, insulin, that is necessary for the proper regulation of blood glucose. But attempts to extract it from the pancreas seemed doomed to fail because a pancreatic digestive enzyme was extracted along with it. Frederic Banting (1891–1941), the Canadian doctor, came up with the idea of tying off the pancreas in dogs and letting the organ atrophy.[19] Extracts of materials from these organs, then, were rich in insulin but lacked the interfering digestive enzyme. Banting's extract proved to be effective on dogs, but was too impure to be used on humans. The person Banting had talked into letting him use research facilities, John Macleod (1876–1935), hired a chemist to do the purification, James Collip (1892–1965).[20] Collip managed to isolate insulin in a pure enough form to be usable on humans, using a method similar to the one Cohn used to isolate serum albumin.[21] Macleod (who did eventually share the Nobel Prize with Banting) started talking about the insulin that "we" extracted and other people started referring to "Collip's extract." Banting, stewing over this one day, sought out Collip and tackled and throttled him.[22]

Though not always so overtly violent, there were other harsh exchanges that occurred in the early investigations of biological materials. The cause for the tensions may be traced to the high stakes that these discoveries were tied to as well as the incredible difficulty of the work, which results, in part, from the fact that many biochemical materials are relatively huge. Water is made up from three atoms. Rubbing alcohol is made from twelve atoms. Hemoglobin is made up from some several thousand atoms. Large biologically active molecules also have a highly convoluted, but highly specific structure. The molecule itself may appear as a jumble of tangled yarn, but each tangle and knot is important to the function of the molecule, as can be demonstrated by the following simple experiment: If a raw egg is placed in a bowl and a teaspoon of salt is poured directly on a spot of egg white, the egg white will appear to "cook"; that is, the clear glob of material surrounding the egg yoke will turn the characteristic white of cooked egg white. (But note: Heat is necessary to kill bacteria in raw egg, so regardless of its appearance, the egg is still unsuitable for consumption.) This change is wrought by the proteins of the egg denaturing, or abandoning their natural shape, in a fluid that is too dissimilar to their natural environment. The forces responsible for the three-dimensional shape of large biomolecules are intermolecular forces, but among them is a type of intermolecular force not previously discussed, hydrogen bonding.

Hydrogen bonding is a type of intermolecular attraction, but it is a particularly strong attraction. Hydrogen bonds occur in materials that are composed of hydrogen and another electron-attracting element, such as oxygen. Water is a prime example, because water has two hydrogens and one oxygen. Alcohol also has a hydrogen–oxygen group, but only one. The strength of hydrogen bonding can be demonstrated by putting equal amounts of water and rubbing alcohol in two different spoons and then tipping the spoons slowly. The alcohol will break loose and flow from the spoon at a smaller angle of tilt than the water because the water is being held to itself in the spoon with twice as many hydrogen bonds. Comparing water with liquid cooking oil, which has no hydrogen bonds, drives the point home.

Dorothy Wrinch (1894–1976), born in Argentina of English parents, grew up in London. She earned a doctor of science degree from Oxford, and in 1935 Wrinch received a research grant from the Rockefeller Foundation with the object of finding mathematical models for biological systems. From her research, she proposed a structure called the cyclol bond. The idea was that proteins are formed from a kind of colloid coagulation, along the lines of Bancroft's description, but with much bigger molecular units forming the colloids.[23] Though her idea was embraced by Irving Langmuir (who was also the supporter of G. N. Lewis's cubic atom for a while) it was rejected by John Bernal. The controversy eventually involved Linus Pauling, who effectively debunked the cyclol bond. Once cornered, Wrinch tenaciously clung to her initial hypothesis, to the eventual detriment to her career, basically spending the rest of her life defending an insupportable concept.

The Braggs and others had a strong interest in discerning the underlying patterns to biological molecules, but they made some erroneous assumptions concerning the many twists and turns, including an underestimation of the importance of the results of quantum chemistry in deciding shapes. As a result they could not arrive at a model that would satisfactorily match the X-ray data. The person who succeeded in sorting out the basic structure underlying the complex tangle of yarn was someone versed in quantum chemistry and X-ray crystallography, someone we have met before: Linus Pauling.

During his career, Pauling tackled an amazing range of topics in physical chemistry and was therefore prepared to take on the problem of the huge biochemically active molecules. But it took him a while to get around to it. By the time Pauling started studying biochemicals, it had been fairly well established that proteins, such as the material of the egg white of our demonstration, are made up of chains of units called amino acids, chemically bonded to each other, end to end. Pauling, working with others, had tried for some time to develop a structure for these chains that would match the available X-ray patterns. In 1948, by his own account, he came upon his breakthrough while he was confined to bed with an infection. Playing with paper cutouts of the chains, he found that if he aligned the hydrogen bonds between the amino acids he could form a corkscrew pattern, called the alpha helix, that would match the requirement of the X-ray data.

In 1954 Linus Pauling would receive the Nobel Prize for his work on molecular structure, especially his work on the structure of proteins. But in 1952 when he requested a passport to discuss his alpha helix with European scientists, the request was denied because the State Department was not convinced that he was sufficiently anti-Communist.[24] From the late 1940s Pauling had been a supporter of several peace organizations and groups concerned with fallout from nuclear weapons testing. In 1963 the Nuclear Test Ban Treaty would be announced on the same day that it was announced that Pauling would receive the Nobel Peace Prize. But in 1952, Pauling couldn't get a passport. Had he received a passport and been able to attend the meeting he wanted to attend, he may have seen Rosalind Franklin's X-ray spectra of DNA. Rosalind Franklin's sharp and detailed X-ray spectra enabled Watson and Crick to propose their famous double helix structure for DNA.

DNA, which stands for deoxyribonucleic acid, has become a household phrase, and most people are comfortable with the notion that DNA is a material found in the cell that carries genetic information. After World War II an intense effort was underway to understand the structure of this important molecule. In 1953 James Watson (b. 1928) had just received his doctorate in zoology from the University of Indiana and was in Europe to study. Francis Crick (b. 1916) was a student still working on his doctorate under Bragg, the

Fig. 31. Linus Pauling, April 28, 1962, outside the White House, protesting U.S. atmospheric nuclear tests. National Archives and Records Administration, photo no. 306-PSA.62-3422, courtesy AIP Emilio Segrè Visual Archives.

elder. Rosalind Franklin (1920–1958) was a physical chemist at King's College who was performing X-ray diffraction studies on DNA. Watson and Crick used the X-ray data of Rosalind Franklin (though sometimes without her knowledge)[25] and models (in the style of Pauling) to successfully propose a structure for this important biological chemical.

But physical chemists have long been interested in the structure and function of biochemical materials. For instance, two of our previous acquaintances among the Ionists, Walther Nernst and Svante Arrhenius, argued bitterly over the application of the law of mass action to toxin–antitoxin systems, but their thinking was linear. Many biochemical systems—the heart, lungs, endocrine system—work with a feedback, switching mechanism: when an enzyme is needed, it is produced; when it is not needed, its production is suppressed. Such systems are nonlinear. They oscillate. Nonlinear dynamics, the study of the peculiar structured and regulated systems that exist in our chaotic world, has become an important research area for physical chemists.

Chapter 29

Nonlinear Dynamics

... our bodies are just wavelets on the surface of the earth ...
Gustav Fechner, circa 1828

Oscillations are a part of life. Swings swing. Waves crash. Old Faithful gushes, and the sun also rises. Oscillating chemical reactions are found—and are necessary—in many biological systems, including contractions of the uterus, the beat of the heart, and biological clocks.[1]

Spontaneous, sustained oscillations can be demonstrated by anyone with carbonated soda and some raisins. If the raisins are soaked overnight in water and then dropped into soda, they will eventually start rolling or rising and falling periodically as the bubbles underneath them accumulate then escape. But this type of oscillation is purely physical, that is, it does not involve a change in the material. What interests us here are purely chemical oscillating systems or those that combine physical and chemical mechanisms.

Oscillating reactions have been recorded in the Western world since the 1600s. Robert Boyle observed oscillating flashes from solid phosphorous in a flask with a loose stopper.[2] The mechanism for this oscillation is both chemical and physical: The phosphorous burns until it runs out of oxygen; then when enough oxygen has diffused back in, it flares again. James Joule observed oscillating electrochemical systems.[3] The type of oscillating system that Joule observed is fairly well understood; when a voltage is applied to an electrode in a reactive solution, such as a corrosive acid, there is a battle between two forces: the push of the applied voltage to grow an oxide layer (such as rust on iron or tarnish on silver) versus the pull of the acid to clean it away. However, oscillations in current-producing cells are a bit more difficult to explain. For instance, one would normally not expect one's AA batteries to suddenly start oscillating. But in 1828, nearly two decades before Joule made his observations, that is what Gustav Fechner saw: oscillations in his battery.

To students of psychology, the name of Gustav Fechner (1801–1887) is familiar. He developed a well-known mathematical model for the response to sensory stimulus known as the Weber–Fechner law, though Fechner insisted on calling it Weber's Law.[4] Fechner's name is less well known to physicists and chemists. This anonymity may stem from the fact that he turned from the study of chemistry and physics to the study of psychology, where he became famous. But it is also partly explained by a propagated bibliographic error. Fechner's papers on his oscillating reactions were published in *Schweigger's Journal for Chemistry and Physics*, a journal founded and edited by Johann Salomo Christoph Schweigger (1779–1857) from 1811 to 1828.[5] However, probably due to space constraints, the journal's name became truncated in some citations to the *Journal of Chemical Physics*, which did not begin publication until 1933. So any requests for Fechner's work would be turned away as an erroneous citation. Such a situation is known as a bibliographic ghost.

Fechner started out his life in a normal manner. He was born near Halle, Germany, in 1801, to a Lutheran preacher and his wife, but his father died when Fechner was five.[6] In 1817 he began attending the University of Leipzig and continued there as student, then professor. He initially studied medicine, but did not practice it, concentrating instead on chemistry and physics. At the age of twenty he started writing a series of social satires using the pseudonym "Dr. Mises." The origin of his nom de plume is not known—Mises was a family name in that area of Europe at that time—but it is also reminiscent of the archaic word *misease*, which means extreme distress or discomfort. In 1828 Gustav Fechner began his experiments with electricity. One day, while measuring the current flowing between a silver wire and iron wire in concentrated nitric acid/copper chloride solution, he saw a strange thing: The current oscillated.

To measure the current oscillations Fechner used a floating-needle galvanometer, which consists of a thin iron wire floating in water in a covered shallow dish.[7] The current-carrying wire from the reaction would run underneath and at right angles to the iron wire. Michael Faraday had demonstrated that electric current induces a magnetic field, which in turn deflects iron wire. The setup made an effective current meter: Every time current flowed, the iron wire would move. In Fechner's experiment it moved right, then left, then right, then left. Fechner, however, did not follow up on his initial discovery, and the exact mechanism for his oscillating reaction is still not completely understood. The reason Fechner did not pursue this further was his other interest: the psychology of sensation.

It was known by that time that electrical discharge to a sense organ could produce sensation. This effect can be readily demonstrated by briefly touching one's tongue to both poles of nine-volt battery: The resulting sensation is a slightly salty taste. These findings are no surprise to us today. The electrical nature of nerve impulses and brain activity is well established, and an instrument has been devised to measure these tiny variations in these voltages, the electroencephalograph. However, to Fechner and his colleagues, these discoveries must have felt as fundamental as Joseph Priestley's isolation of oxygen.

Fechner, like Loeb, was very interested in the connection between physicochemical systems and how the mind works. Unfortunately, it was this interest in sensation that caused his catastrophic decline in health, both physical and mental. In the course of pursuing his interest in the physiological origin of sensation, Fechner became fascinated with afterimages, the dark images in the eyes after exposure to bright light, such as the spots before the eyes after the flash of a flashbulb. These visual afterimages, which are now regarded as little more than a mildly entertaining optical illusion, became a focus of interest for Fechner. Unfortunately, in 1840, he went so far as to gaze at the sun to produce the effect. Although he used colored glasses during the experiment, he went blind. The stress of the blindness, in turn, seems to have caused Fechner to suffer a nervous breakdown, which resulted in his resignation as chair of physics at Leipzig.

After three years of blindness, while out in his garden, he removed his bandages and found he could see—in fact he could see so well he was convinced that he could see the souls of plants. Fechner seemed to struggle with his own madness the way a physician might struggle with his own cancer. Although at first he was forced to be qualitative in his efforts to understand the workings of the human mind, he was no doubt nagged, by his training in physics and chemistry, with a desire to be more quantitative. His wish was fulfilled when he hit on the idea for a mathematical model that would relate the increase of physical stimulation to the increase of the corresponding mental sensation.[8]

Fechner's finding was influenced by the observation made by Ernst Heinrich Weber (1795–1878) that the change of stimulus necessary to be noticeable increases as the stimulus increases—which might be summarized by saying "the more you have, the more it takes to notice a difference."[9] An analogy might be found in weight loss: If a 300-pound person loses ten pounds, the difference is not likely to be noticed. If a 100-pound person loses ten pounds, the change is profound. Weber said this was true for sensation, too, and he demonstrated his rule by having subjects judge differences in weights, line lengths, and some tactile sensations.[10] The reader may confirm the effect by a simple experiment with pennies. A subject, sitting with eyes closed, is handed the two containers with pennies and asked to judge if they contain the same or a different number of pennies. The greater the number of pennies, the greater the relative difference necessary before the subject can make the distinction. In other words, if the subject is trying to decide between one penny and two pennies (one more penny), the difference is obvious. However,

Fig. 32. Gustav Fechner. Archives of the History of American Psychology, Photograph File, The University of Akron.

the difference between forty pennies and forty-one pennies (again, one more penny) is difficult to detect.

Although Fechner relied on subjective judgment of sensation to validate his relationship, it has been subsequently shown, using electrophysiological measurements of response, that the relationship really does follow Fechner's mathematical function over a limited range of stimulation. In 1860 Fechner published his mathematical formulation of Weber's law in *Elements of Psychophysics*, which is considered the first publication of that discipline. The importance of Fechner's contribution to the field of psychology—finding a psychological phenomena that could be modeled with mathematics—is evidenced by references to Fechner's law in standard textbooks in psychology and Freud's inclusion of Fechner's insights into his 1900 *The Interpretation of Dreams*.[11]

Gustav Fechner, true to his age, was a virtuoso of science, interested in all problems to which he might apply his skills. As a result he made significant contributions to psychology as well as electrochemistry. He received substantial accolades for his foundation of psychophysics, but he would have been disturbed by the neglect of his contributions to chemistry and physics. Hopefully we have laid this bibliographic ghost to rest.

Fechner's oscillating reaction is a heterogeneous system; that is, it takes place in two phases: the solid-phase electrode and the liquid-phase solution. In the early 1900s other heterogeneous oscillating systems, such as periodic dissolution of iron wire in nitric acid,[12] were discovered. Though reported in the literature, these reactions caused little stir. Since the work of Helmholtz and Gibbs, it had become chemical dogma that all well-mixed chemical reactions proceed smoothly toward equilibrium, not in fits and starts. Saying that a homogenous, well-stirred, solution-phase chemical reaction could oscillate was like saying that a bowling ball might spontaneously start jumping up and down. In 1921, William C. Bray, an associate of G. N. Lewis originally from Canada, found that hydrogen peroxide could be made to decompose periodically if a catalyst were added. Anyone who has used drugstore hydrogen peroxide as a disinfectant has seen the way it spontaneously bubbles. When the proper catalyst is added to concentrated hydrogen peroxide, the bubbling comes in periodic bursts that can be as regular as a heartbeat. Though this reaction occurred entirely in the solution phase, the report received a cool reception. The reaction was written off as an experimental artifact caused by an impurity. This skepticism of his results and questioning of his methods must have been highly offensive to the taciturn Canadian. Though the reaction has subsequently been shown to be a true homogenous oscillating reaction, Bray does not seem to have pressed this line of research. In 1955 Ilya Prigogine demonstrated theoretically that oscillations can exist in homogenous systems where the reactant and product concentrations are greatly different from what they will be at equilibrium, a condition known as "far from equilibrium."

Ilya Prigogine (b. 1917) a physical chemist, was born in Moscow a few months before the revolution of that year. His father, a chemical engineer, took his family to Germany after the revolution and then Belgium. Prigogine did not originally intend to study chemistry. His interests lay more in history, archaeology, and music (he reportedly could read music before he could read words). But physical chemistry won him over because of its strong link between chemistry and mathematical theory. The basic idea underlying Prigogine's model is well represented by the analogy with rabbits and wolves. In a lush field of grass, a pair of rabbits will blossom into a troop of rabbits. As long as the fields stay lush, the rabbits will multiply. Eventually a large rabbit population will attract wolves. When the wolves move in, they have plenty of rabbits to eat, so they will multiply. Eventually the wolves will become so populous that rabbit repopulation will not be able to keep up, and the rabbit count will start to go down. When the rabbit population drops, it will reach a point where the wolves find themselves strapped for food and the wolf population will start to drop, too. When the wolf population reaches a low, the rabbits will start to rebound. And so it goes. In the model, the situation is held "far from equilibrium" by the fact that there is a lush field of grass: a continued supply of food keeps the system from running down. If the grass ran out, the wolves would win for the moment, then they would die out, too.

Strong objections were raised to this concept. How could oscillations be sustained if the basic laws of thermodynamics were to be obeyed? Didn't all reactions have to proceed toward the maximum evolution of heat and the maximum entropy? Prigogine pointed out, once again, it was the net of these two effects—the free energy—that was important. While the free energy did have to proceed downhill toward equilibrium, there was nothing to prevent the heat, the entropy, or even concentrations from oscillating along the way. In the systems he described, there was plenty of opportunity to do this because they were far from equilibrium, just as the rabbits and wolves are held far from equilibrium by the lush fields. As long as there is a supply of grass, the cycle can continue. If there is a drought and no more grass, the system runs down. But Prigogine needed an acceptable real-life example to substantiate his ideas. Prigogine won the Nobel Prize in 1977 for his work, but he might not have if it weren't for an obscure fellow Russian by the name of Boris Belousov. In the 1950s a student of Prigogine came back from a visit in Russia to report a reaction that matched Prigogine's model.

Boris P. Belousov (1893–1970) spent his youth in czarist Russia, though after the revolution of 1905 his family left for Switzerland, were he studied chemistry. As a young man Belousov returned to Russia. From there, the details of his life remain sketchy. Belousov discovered his oscillating reaction around 1951, but was unable to get an article on it published because no reputable journal would believe that such a reaction were possible. When enough people heard about it by word of mouth, he was urged to find a venue for publication just to establish his priority of discovery. In 1959 he finally did: He slipped it into an in-house publication issued by the institute for which he was working. The reaction appeared in a booklet entitled *A Collection of Short Papers on Radiation Medicine*.[13]

The reaction is now known affectionately among physical chemists as the Belousov–Zhabotinsky or BZ reaction. The B is for Belousov and the Z is for Anatol Zhabotinsky, the chemist who started its systematic investigation. It is a beautiful reaction to witness: The stirred reaction solution cycles periodically between rich red, deep blue, and new-leaf green in a series of sudden and dramatic color changes that can be made to repeat for thirty minutes or more. Recipes for the reaction can be found in many handbooks for chemical demonstrations,[14] and the reaction itself has been widely investigated since its popular reception. But initially it was known only by one person telling another. One of the people who received the information was Anatol Zhabotinsky (b. 1938).

Zhabotinsky, a graduate of the biophysics department of Moscow State University, was intrigued by reports of oscillations in glycolysis. Glycolysis is the series of reactions in the body in which sugar is broken down for energy. However, while trying to assemble the materials to study this reaction Zhabotinsky heard about Belousov's solution. When he tried it and saw the beautiful pulses of color, he wrote enthusiastically to Belousov of his desire to study the reaction and asked to meet with him. But Belousov had retired by that time and pleasantly but firmly asked not to be disturbed.

Zhabotinsky continued to study and publish on the reaction. Eventually knowledge of it spread outside Russia and to the United States. In the United States, a distant cousin of Arthur Amos Noyes, one of the heirs of the Ionists, Richard Macy Noyes (1919–1997), heard of the reaction. It soon became the passion of his life, too.

Richard Noyes's father, William Albert Noyes (1857–1941), chairman of the department of chemistry at the University of Illinois, was sixty-one years old when Richard Noyes was born.[15] The story is told that the dean hinted that the elder Noyes should think about retiring, at which point the dean was invited to attend the christening of Richard's younger brother.

Not only was Noyes's father a chemist, but so was his half-brother. When Richard Noyes entered Harvard he majored in chemistry because, by his own account, he had no imagination. His experience at Harvard was actually his first formal training in chemistry; his father had advised him against taking chemistry in high school. Noyes graduated summa cum laude. Though he had the opportunity to work with Linus Pauling in graduate school, he decided not to because he was more interested in chemical processes than the structure of molecules.

Noyes married a chemist, Winninette Arnold. Their early life together was marred by sadness. His wife suffered from diabetes and two of their children died in infancy. In addition he chose some unprofitable areas to investigate, and after a thirty-year career, it did not appear that Richard Noyes was going to make much more than a prosaic contribution to science.

Then, in 1969, he learned of the Belousov–Zhabotinsky reaction from a colleague returning from work with Prigogine. Noyes's many years of study then came to fruition. By 1971 he had found he could account for the observed oscillations by a reaction mechanism that involved a series of eighteen separate steps and twenty-one different chemical species, including the transition states. In other words, our rabbits and wolves are a highly simplified model; Noyes found he had to consider the entire food chain. He offered a simplified model of five idealized steps that exhibited oscillations and called it the Oregonator (he was living in Oregon at the time). For the next twenty-five years he wrote over 104 publications on the BZ and other oscillating chemical reactions, basically establishing this area of research in physical chemistry. All of this work was accomplished after his fiftieth birthday.

Originally, the BZ reaction was studied as oscillations over time, but Prigogine, Noyes, and Zhabotinsky were also interested in reactions that changed as a function of distance. These reactions are called spatial oscillations, or waves. If the BZ reaction is initiated in a shallow dish that is not stirred, the reaction will start with a blue spot that will expand outward. After it has expanded some distance, the center where it started will turn red. This pattern of reaction and diffusion continues until there is a red and blue design of squiggles and swirls.

Other reactions also display pattern-forming behavior.[16] The curious nature of spontaneous pattern formation and organized structure interested the mathematician Alan Turing. Turing (1912–1954) was a British mathematician who did important work in computer design and who was instrumental in deciphering German encrypted messages in World War II, work for which he was awarded the Order of the British Empire.[17] In 1952 in a landmark paper in theoretical biology, Turing presented a mathematical model for a chemical mechanism that could cause patterns such as those seen in the stripes of zebras or a leopard's spots.[18] The mechanism incorporates reaction followed by diffusion, such as in the spatial waves in the

BZ reaction. Though the actual mechanism for natural patterning is now thought to be more complicated, Turing presented the foundation on which further studies would build.

But further work would have to be done by others. In 1952 Turing called the police to report a robbery of his home. While the police were there to investigate the robbery they also found evidence that Turing was homosexual. Turing, an academic who was fairly naive, had never taken pains to cover up his personal life, apparently not believing that it was necessary. However, an 1885 law was invoked to arrest Turing and he was subsequently tried and found guilty of gross indecency. Although fellow scholars and colleagues spoke up for him, he was sentenced to undergo drug therapy meant to suppress his homosexuality. He survived a year of drug treatment and a year after that, but in 1954 he was found by his housekeeper, dead of cyanide poisoning. There was no note.[19] He was forty-two years old.

Turing would have been gratified to know that in addition to the fascinating zebra-like patterns that emerge from a perfectly uniform BZ solution, there are also structures that can develop on a scale indiscernible to the eye. With the advent of new microscopy instrumentation we can now image surfaces on the billionth of a meter scale.[20] Sometimes microscopy reveals the rough and erratic mountains that we would expect to see, but sometimes we see delightful patterns of honeycombed or gridded surfaces. These minuscule structures are said to be on the *nanoscale*. A nanometer, a billionth of a meter, is the standard unit in this strange, diminutive world.

Chapter 30

Nanotechnology

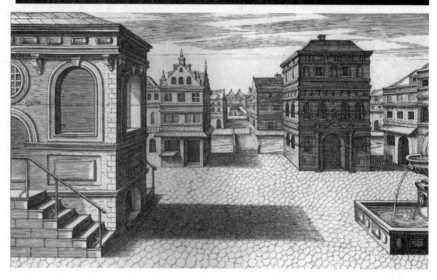

There's plenty of room at the bottom.

Richard P. Feynman, 1959

The U.S. military once tested Richard Feynman's IQ and rated it sub-standard, as Feynman likely intended it to be.[1] Richard Feynman (1918–1988) had a good sense of humor and went on to share the Nobel Prize anyway, for his work on quantum electrodynamics. In 1959, before he won the prize, already a well-respected physicist, he was invited to give a talk at the American Physical Society. He decided to make the topic of talk rather speculative, but he didn't talk big. He talked small. He spoke about the possibility of building motors tenths of a millimeter on a side, writing libraries on the head of a pin, and contriving machines that could be swallowed to perform microsurgery.

Feynman's idea captured the imagination of many, including a name that has become associated with the nanoworld—the world on the billionth of a meter scale—Eric Drexler (b. 1956). Drexler has made a career out of predicting the future of nanotechnology. He, with his wife, Chris Peterson, a chemist, founded the Foresight Institute in Palo Alto California in 1986 where they sponsor seminars and write on the possibilities of nanotechnology.[2] Drexler's predictions are even more ambitious that Feynman's. Drexler envisions a world where nanoscale robots manipulate individual molecules and atoms, positioning them with tiny tools, building minuscule machines, and assembling replicas of themselves. Though Drexler's specific forecasts have come under considerable criticism,[3] nanotechnology is currently an exciting area of research in physical chemistry, though not exactly in the fashion Feynman imagined.

To begin with, we need to excuse Feynman from the fantastic because Feynman was still talking about scaled down but microscopic, not subsubmicroscopic devices. Feynman issued a challenge to make a motor a tenth of a millimeter on a side, a challenge that was quickly met. But the millimeter scale is still 10,000 times bigger than such a motor on the nanometer scale. A nanometer is about the diameter of one molecule of sugar.[4] A nanometer compares to a meter as a meter compares to the distance to the moon. On the nanometer scale, systems behave differently from what we are accustomed to. To use what has become a trade witticism, size matters.

For instance, if you scale down an airplane to nanoscale dimensions, it could no longer fly. It would be bombarded by air molecules as big as cannonballs and there would be nothing but vacuum between collisions. And you wouldn't be able to machine parts to fit. You can build only with integral numbers of atoms. You couldn't shave off a quarter of an atom here or an eighth of an atom there. And there would be those pesky intermolecular forces to deal with. It wouldn't be simply a matter of picking up an atom and placing it where it needs to be. You would have to find a way to shake it off, too. The intermolecular attractions would make it more like trying to stack magnets with steel pliers than laying bricks.[5]

The difficulties mount. Properties of materials are different on the nanoscale. To understand why this might be, consider a large group of people carrying a twelve-foot canoe over their heads. A

Fig. 33. Richard Phillips Feynman with Toichiro Kinoshita, June 1961. Clarice Schwinger, courtesy AIP Emilio Segrè Visual Archives.

group of people can do this, but one person cannot. But the canoe-carrying properties of the group don't decrease gradually as the number of people in the group is reduced: Removing one person from the group doesn't just lower the canoe a couple of inches and removing another person lower it a few inches more. There is some critical bulk number of people necessary to carry a canoe, and when there aren't enough people, the canoe drops to the ground. The analogy carries over to the nanoscale. Nanoscale droplets of water, known as nanowater, do not act like bulk water until some 5,000 molecules have accumulated. Because there are so few molecules in the structure, when nanoaggregates are hit with a heat source they don't melt from the outside in, the interior starts melting at the same time as the outside. Nanoparticles can be invisible, in the literal sense: not only hard to see, they can be transparent, too. Sunscreen is made of nanoparticles of titanium oxide, the stuff that used to look like a white paste on sunbather's noses when sunscreen was non-nanosized. You may have noticed there haven't been a lot of white noses at the beach lately. Researchers are also looking at ways to incorporate nanoparticles into a paint that would provide a visible-light-rejecting sunscreen for cars. Paint would last longer and the interior would remain cooler. So an entirely different mode of thinking has to be adopted for engineering on the nanoscale. But somebody must believe it is possible. There is mega-financing for nanoscience. Research is currently being funded to the tune of hundreds of millions of dollars per year.[6]

Hype and hyperbole aside, something is already working on the nanoscale. Nature is. There is no more impressive, entropy-defying, energy-utilizing, self-assembling, self-repairing, environment-responding nanorobot than the human cell. Many researchers are looking to use biomolecules to do a lot of the work for them. In this way, nanotechnology may just be an old concept with a new name. The idea, of course, is to do it a bit better. High on the nano wish-list are things like nanodevices to detect cancer when it is still only a few cells wide and devices to deliver drugs directly and exclusively to the site.[7] It has been proposed that a virus, emptied of its infectious contents, would make an excellent vehicle for tiny drug deliveries. Viruses naturally target specific cells. In addition, the viruses reproduce readily, so they could be farmed for a continued supply of nanovessels.

Some progress has been made in not-so-natural nanogizmos, too. A nanobalance has been devised that can determine the mass of nanoparticles. How does a nanobalance work? The frequency with which a fiber vibrates is proportional to its mass. When as little as a quadrillionth of a gram is absorbed on these tiny balances the change in resonance frequency can be detected by seeing what frequency of light makes it vibrate. Such a device, with a selective molecule attached, could make a single-molecule sensor.[8] For instance, an enzyme that binds with sucrose could be attached to the fiber and when one molecule of sucrose finds the enzyme, the vibrational frequency of the fiber would change. A similar technique might be adapted to find a single molecule of a toxin or a defective strand of DNA. Methods for manipulating atoms with macroscale machinery have been contrived, as evidenced by the famous IBM logo mentioned previously that was assembled from xenon atoms.[9] Atomic switches have been devised: minuscule structures called quantum dots have been made to trap single electrons and could be made to function as nanotransistors in miniature computers.[10]

However, there are lots of items on the nano wish-list. Nanogears and nanowrenches are needed for the nano toolbox. Nanopumps and nanomotors are needed for active devices. Some ingenious ideas have been put forward for these last two. The biological world, again, provides a model for a nanopump to emulate. In living cells there can be a flow of material against the concentration gradient; that is, there can be material flowing from areas of low concentration to areas of high concentration (like water running uphill). To effect this flow for water, a pump would be required; to effect this flow in cells, similar devices are needed, and nature has found them. The best example is an ion pump. This biological structure pushes material in one direction only and against the concentration gradient. A nanopump could be designed similarly. It could have an entrance that randomly opened and closed but only one exit. Net movement would be in one direction only. Such a pump could send material against the entropy gradient from areas of low to high concentration: Maxwell's demon come to life.

A nanomotor would be tricky to devise, but there are some interesting ideas in the works there, too.[11] The problem with nanomotors is that they cannot operate on macroscopic gas expansion like internal combustion engines or relatively massive electron flows like electric motors. These tiny machines would be just slightly bigger than the gas particles themselves and would constantly be bombarded from all sides at once. But there is one way the motor could take advantage of the random bombardment: by ratcheting.

To understand how this works, consider bathing babies in a sink versus a bathtub. In the bathtub they splash and play randomly and send the water in all directions, but there is no net movement of water. In the sink they splash and play randomly, too, yet there is a net movement: The water moves out of the sink and onto the floor. The difference is that the bathtub water, temporarily displaced, can roll right back because of the relatively high sides on the bathtub. In the sink, the water goes out, but can't come back in. So the random motions of the babies in the bathtub don't cause a net displacement of water, but those of the babies in the sink do. Ratchets in a turnstile work on the same principle: They turn in one direction but won't turn in the other direction. Researchers have devised a Y-shaped molecule and a G-shaped molecule that they managed to fit together in a turnstile-type arrangement where the Y molecule allows the G molecule to turn in one direction only by jamming motion in the reverse direction. Many of these proposed products will incorporate a structural component that may be nominated the nanobrick in the nano arsenal: buckminsterfullerine. Bucky balls.

One of the first triumphs of X-ray crystallography was J. D. Bernal's solution to the structure of graphite, which turned out to be sheets of carbon that can slide over one another. One of the first triumphs of nanotechology was to identify a new form of carbon, a sixty-carbon ball that has remarkable resemblance to a soccer ball and has been named buckminsterfullerene. Buckminister Fuller, of course, was the architect who incorporated the geodesic dome in his designs, which buckminsterfullerene resembles. This name was quickly shortened to "bucky balls." At the time they were discovered, however, the discovers didn't know they were contributing to nanotechnology because "nanotechnology" had yet to come into the common parlance of physical chemistry.

Richard Smalley (b. 1943) (a last name that begs comment in our discussion of nanoparticles) was using a laser to blast a few atoms at a time from a solid to make small-number clusters of atoms, on the order of ten to twenty atoms per clump. At the suggestion of Robert Curl (b. 1933), Harry Kroto (b. 1939) visited Smalley's lab to find out if long-chain carbon molecules might be made in carbon-rich stars.[12] Carbon's ability to form long chains and complex structures by binding to itself accounts for its prevalence in living materials. They found the mass of their clusters using a mass spectrometer—an instrument built on the idea that J. J. Thomson and his colleagues used to find the charge-to-mass ratio of an electron. That is, Thomson bent a charged beam with a magnet. The curvature will depend on both the strength of the magnet and on the mass of the particle in the beam. Thus, through this method, after they knocked one or more electrons off their clusters, they were able to measure the mass. They found that the mass spectrum was definitely dominated by one particular mass: a mass that corresponded to a particle made up from sixty carbon atoms.[13] They made a model of the molecule and decided it resembled a geodesic dome. They won the 1996 Nobel Prize for their efforts.

Then in 1991, as Sumio Iijima was inspecting pure carbon soot with an electron microscope in a laboratory in Japan, he noticed there seemed to be odd-looking threads running through the material. On closer inspection, he found them to be long hollow symmetric tubes now known as *nanotubes*. These nanotubes may eventually be the nanoframework for nanostructures or the nanowire for nanocomputers. The intermolecular forces between multiwalled nanotubes, working as a temporary attraction rather than a permanent chemical bond, may allow such nanotubes to act as nanobearings.[14]

Nanotubes are amazingly strong structures: Pound-per-pound they are stronger than steel but pound-per-volume they are one-sixth the weight. They also have been found to have extraordinary tensile strength and thermal stability, which has led to speculation on their use in the macroscale world as well. Suggested ideas have included earthquake-resistant structures and car bodies that absorb shock and bounce back from dents. The nanotubes conduct heat and electricity better than copper. A paint that incorporates nanotubes has already been developed that can support a static charge and therefore make a better bond to the surface being painted.[15]

But for all the wonder and dreams of possibility, a moment's pause must be taken to consider Eric Drexler's second thoughts. Drexler introduced into our thinking what he has called the "gray goo problem." If we have skin-cell sized robots that can self-replicate, what's to stop them from self-replicating uncontrollably? Should we panic and put the skids on all this nanotechnology research? What are the possibilities? What are the limitations? To gain an idea of the nature of the answer to this last question, let's take a glance at another area of physical chemistry research that is of great interest to Smalley: molecular beams.

For those who remember the Star Wars initiative of the Reagan presidency, the term "molecular beam" might be familiar. Yuan T. Lee did epoch-making research with molecular beams and shared the 1986 Nobel Prize with John Polanyi and Dudley Herschbach. Although molecular beams have some definite limitations as satellite-based weapons (a lightning-bolt molecular beam in space would have diverged to a spring shower by the time it reached Earth), they are very effective tools for studying chemical reactions on a particle-by-particle basis; they remove the confounding effects of multiple collisions or solution interference. When material is put into a beam (which can be accomplished by allowing it to expand through a small orifice), all the molecules are basically traveling in the same direction and there are far fewer intermolecular collisions. The first molecular beam work was done in the United States in the late 1920s by Harold Urey and Francis Owen Rice. By using molecular beams they were able to provide the first truly convincing evidence against the radiation hypothesis. They showed that the extent of reaction depended on the number of collisions, not how much light was shining on the material.

Richard Barry Bernstein (1923–1990), was the son of Russian immigrants. While working toward his Ph.D., he joined the Manhattan Project. He decided to remain with the army and ended up being part of the first Bikini Atoll atomic weapons testing. The experience was an interesting one for him because he and a crew were sent to do a last check of the instruments before the test and got stranded on the island. When it became apparent that no one had noticed their absence they spelled out "SOS" on the beach with strips of underwear. When that didn't work they sent back falsely high radiation readings on the instruments they had been sent to check. This last one got the attention they needed and they were lifted off the island.[16] Bernstein helped to develop an extensive body of theory for molecular beam studies.

What do molecular beams have to do with the possibilities and limitations of nanotechnology? Two of the effects that Bernstein observed in molecular beam studies, rather poetically called rainbow scattering and the glory effect (or Bernstein's wiggles),[17] are actually quantum effects caused by interference. Atoms causing interference patterns? Aren't atoms big with respect to electrons? When Richard Smalley's sixty-carbon bucky balls are put in a molecular beam apparatus, they create interference patterns, too: spots of dark where the material was and spots of clear where the material was not.[18] Constructive and destructive interference. Wave behavior.

There is something about bucky balls behaving as waves that offends the sensibilities. We have become used to the idea of sound waves acting as particles and light particles acting as waves, but sound and light are things we can turn on and off. How can you turn on and off a particle of soot? Something as big as a bucky ball is just not something you expect to blip in and out of existence. An electron maybe. It's small and moving at a significant fraction of the speed of light. But a bucky ball?

These are the considerations that pose limitations to how nanotechnology can ever emulate macroscale engineering. At the atomic level there is a physics that has to be acknowledged and dealt with. In technical parlance, these are called "weird quantum effects," and they will determine the possibilities of nanotechnology.

Chapter 31
Extreme Quantum

Anyone who is not shocked by quantum theory has not understood it.
Niels Bohr, circa 1930

O ne thing that must be clearly stated before we dive into any criticism of quantum mechanics is that quantum mechanics has been very good to us. Quantum mechanics provides the basis for spectroscopy, the single most powerful analytical tool to date. It has made possible imaging techniques used for noninvasive diagnosis in hospitals. Quantum mechanics has given us the semiconductors and lasers that have advanced technology from surgery to weaponry, as well as information storage and retrieval.[1] Warts and all, it's still a pretty good system.

Researchers have dealt with many of its bugs or worked around them. Among the major triumphs of quantum mechanics in recent years has been the discovery of mathematical algorithms to simplify the daunting equations and the use of computers to crank through the intense calculations. In 1998 the Nobel Prize went to Walter Kohn (b. 1923) and John Pople (b. 1925) for accomplishing just this. Walter Kohn developed *density-functional theory,* a mathematical model that reduces the complexity of many-electron problems by using a sophisticated averaging method. John Pople developed computer methods to make the computations possible.

Walter Kohn was born in Vienna, Austria, in 1923 and by his own account, genius was not foreshadowed in his life.[2] Initially, Kohn received only average grades in mathematics and did not display exceptional interest in science. However, when Germany annexed Austria, his parents managed to get Kohn, his sister, and brother out of Austria. In England, his guardians encouraged him to go to school where he studied mathematics, chemistry, and physics. When he was seventeen, because of his Austrian citizenship, he was interned as a foreign national of an enemy state. But the school he had been attending continued to send him books. Scientists, likewise interned for their "enemy alien" status, set up impromptu lecture series, a fortunate event in the midst of misfortune. This haphazard education was continued when he was shipped with other internees to Canada. In 1942, at the age of nineteen, he was released.

At that point Kohn planned to become an engineer so he could take care of his parents after the war, still believing them to be alive. But he was convinced to study physics instead, being assured that he could make a living in this field as well. Special dispensation was procured so that he could proceed in this program without the benefit of a chemistry course because, as a German national, he was not allowed into the chemistry building. From there on, his life resumed a more regular course, excluding the fact that his research eventually culminated in the Nobel Prize. In 1963, on achieving what was to become density-functional theory, he recalls that the results were so extraordinary that at first he did not believe them himself.

John A. Pople was born in England in 1925, to a father who was the owner of a men's clothing store and a mother who was a sometime tutor and librarian. He was the first in his family, to his knowledge, to attend a university. He credits his parents' ambitions for him for his subsequent success at school, despite having to attend classes in the midst of air attacks. When he was twelve years old, he became very interested in mathematics but, probably thinking it too impractical, hid the fact from his parents. He completed his own school books on the subject and then found some discarded textbooks for further reading. Taking things to an extreme that would occur only to a twelve-year-old, he actually hid his preoccupation by purposely introducing errors into his mathematics classwork.[3] However, his joy in his accomplishments eventually became too difficult to suppress, and when the school realized his talent, his education in mathematics began in earnest.

These two men formulated powerful methods that are used to model large, intricate molecules and complex chemical interactions. And they accomplished their results using the "shut up and calculate"[4] approach, that is, ignoring all the quantum weirdness. The reason they, and the rest of us, are able to proceed with our lives despite the nagging weirdness is probably because there are really two categories of quantum peculiarities. One is a weirdness that has been accepted for so long that we have come to believe it or for which there are enough real-world analogs that it can be dressed up to be palatable. The other we'll address later.

In the first category we can put tunneling, spin, and the wave–particle duality. The first of these, tunneling, is the weird process on which the scanning tunneling microscope functions. As may be recalled, *tunneling* is the ability of an electron to be in a region that in classical physics is forbidden, such as a baseball penetrating a brick wall. We accounted for tunneling by saying that the wave function can be interpreted as a probability wave, and even though there is no probability that a baseball will go through a brick wall, there is some finite probability that the electron will tunnel out of the tip of a stylus that is being scanned over a surface. Tunneling doesn't seem completely bizarre if you accept the electron wave as a probability wave.

Spin, likewise, can be fathomed. We have already said electrons aren't really "spinning." Paul Dirac showed that spin is a consequence of Einstein's relativity and arises naturally from the Schrödinger wave equation when it is acknowledged that electrons can travel at close to the speed of light. But a spinning top is a good model for this property of electrons and it is a great temptation to start believing that the electrons must really be spinning. However, recall that a sound wave can shatter glass so it has a virtual mass and real velocity, so the mathematics that model a bullet flying through air can also be used to model a sound wave—but sound isn't bullets and electrons aren't really spinning. We call it spin because that is a concept we can envision. We accept that it is a metaphor and proceed.

The spin model can be used to calculate energy levels that agree with spectroscopic measurements and thus has gained some legitimacy.

The wave–particle duality of light may take a bit longer to accept, but again isn't completely unpalatable. We said that sound also acts as a wave when it bends around corners and behaves like a particle when it shatters glass. We explained interference patterns created by light waves in terms of crests and troughs in water waves and with combs. Interference patterns formed from electrons and bucky balls are a bit stranger (and are probably what prompted Feynman to call interference "the mother of all quantum effects").[5] But we can replicate interference patterns in our own living room, if we like, with a laser pointer and metal ruler, as we described in our discussion of electromagnetism, and this seems to lend credence to the concept.

Another manifestation of the wave nature of matter that scientists have accepted is the existence of Bose–Einstein condensates. This new phase of matter owes its existence to spin. We have said that two electrons cannot occupy the same place in space at the same time (which makes sense), but it turns out this is only because they have a half unit of spin. Particles that have a whole unit of spin, such as photons, turn out to have an interesting property: There is nothing that prevents them from coming together and moving as one. In the quantum world, sometimes you *can* park your car where your neighbors have already parked their car.

The person who came up with this idea was Satyendranath Bose (1894–1974), who was educated at the University of Calcutta and then taught there, with one interruption to teach at another Indian university, until retiring in 1956. Bose recognized that photons have an integral spin (they have a spin of one). He used this fact and a statistical approach similar to the one used by Maxwell and Boltzmann to reproduce the skewed curve of blackbody radiation that Planck had wrestled with.[6] He submitted his work to a respected English journal for publication, but they rejected it. He then sent the paper to Einstein, who appreciated the work, translated it into German, and submitted it to a respected German journal with his recommendation. They published it.

Einstein, being Einstein, took the matter one step further. He said the theory was not only true for photons, but for any particle with an integral spin, which included many fairly substantial atoms. The tantalizing consequence was that these particles should be able to condense if they were cooled enough—but not condense in the manner of a gas condensing to liquid, but condense right into one another. The matter waves should overlap. They would no longer be separate particles. They would share the same quantum state and move together in unison.

A protest, as expected, arose. Some said that it was a reach to apply mathematics meant for photons directly to atoms. Some said that Einstein had done the math wrong. But Einstein and Bose would have been gratified to know that in 1995 Eric A. Cornell and Carl E. Wieman managed to achieve a condensate of 2,000 rubidium atoms that lasted for ten seconds after they cooled it to a tenth of a millionth of a degree above absolute zero.[7] Once again, although we may not like it, we seem to accept it. After all, laser light is produced by many waves of light moving together in the same state of motion. Superconductivity is now explained in terms of coherent motion of pairs of electrons.

However, there is another class of weirdness, alluded to above, that has no classical pattern to explain it. First there is a phenomenon called superposition. When a measurement of a quantum system could have one of two possible outcomes, the system exists as a superposition of those two states until the measurement is made, and then the system suddenly "becomes" just one state—the state that is measured. In the parlance of the trade, this is called "collapsing into the measured state." There is no need to make up an analogy to explain this idea; Erwin Schrödinger already did it for us. Shrödinger's Cat is one of the most famous analogies in quantum mechanics.

Schrödinger said to consider the following: For an hour, a cat is placed in a box with a vial of poison gas, a radiation detector, and a radiation source that has a fifty–fifty probability of emitting radiation in that hour. If the source does emit radiation then the detector will detect it, and circuitry and machinery hooked to the detector will open the vial and kill the cat. So in an hour's time the cat has a fifty–fifty chance of ending up dead. (For cat lovers out there, please accept the appropriate apologies. The cat was Schrödinger's choice.) For that hour that we are waiting, the cat is in a superposition of states: dead and alive. There is no way to tell if the cat is dead or alive except to look in the box. As soon as someone looks, the situation collapses into one solution: dead or alive. Initially, this may not appear too weird. After all, the outcome of a coin toss is not known until the coin stops. Until that time it could be said that the coin is both heads and tails.

But it gets weirder. There is also a property of quantum mechanical systems called *entanglement*. Entanglement means that quantum particles can be so intimately connected that when a measurement is made for one, the outcome for both is determined. Measuring one predicts the state of the other. Again, this may not appear all that strange on the surface. When the coin is determined to be heads-up, that assures us that is it also tails-down. But for quantum mechanical systems there appears to be some evidence that this entanglement may be true even if the particles are separated in space. It's as if one coin landing in New York flipped another coin in Los Angeles. Or looking in at Schrödinger's cat and finding it alive immediately killed another cat in another box somewhere else. Cat lover or not—that's weird.

There have been experiments done with particles (leaving the cats alone) that seem to confirm the effect. To understand the experiments, first consider the different possible states of spin. If we mentally paint our particle orange on top and white on the bottom then we can look at it and say if it is spinning top-up or top-down. It can also be spinning left or right. Taking all these together, then, we have four possible combinations: up versus down and left versus right. If we measured the spin of a stream of electrons from an arbitrary source (such as Thomson's cathode rays), we would find half with top up and half with top down. We would also find half spinning left and half spinning right. This seems right and it probably would have made Emmy Noether feel good, too, because it agrees with her finding that there should be a conservation of symmetry.

Now, in the experiment, a beam of right-spinning electrons is selected from a random beam. These right-spinning electrons are then divided into spin-up and spin-down beams. When these up–down beams are recombined, the resultant electrons are just as they were before the up–down split: spinning right. But when the spin-up beam is blocked and not allowed to recombine with the spin-down beam, the spin-down beam can be analyzed and found to have both left- and right-spinning electrons. At first we may feel relief—some down spins have turned to up spins, flipping from right to left in the process, and symmetry is preserved—but at what cost?[8] How did the electrons *know* that the other beam was blocked and that they ought to switch their spins to half and half?

If all this spinning is getting confusing, consider the following model for a similar situation. Say we have a jacket zipper that opens at the bottom. Now let's paint one side of the zipper red and the other side green. When we zip the zipper up we will have red and green interlocked teeth, colors changing every other tooth. But now consider unzipping the zipper, throwing one half away, and running the zipper up the side that is left, say the red side. You would not expect to look down and see that this side of the zipper had suddenly changed from all red teeth to a pattern of red and green teeth, colors changing every other tooth. But this is what the electrons did when they switched to half left and half right spins. To restore its symmetry this way, the one zipper side would somehow have to *know* that the other side was missing.[9] It is *knowing* the fate of the other particle that is the difficult-to-rationalize aspect of superposition and entanglement.

Evidence seems to show, however, that this is what happens, though some have questioned the experimental methods.[10] Of course this is not the first time quantum mechanics has faced a conundrum. Neils Bohr's model was used to explain quantized energy levels although its orbits should have instantly destroyed the atom it modeled. Samuel Goudsmit talked about electron spin, but Dirac showed that electrons could not actually be spinning. Quantum mechanics has a history of insupportable hypotheses, but you have to start somewhere. The keyword is "start." Quantum mechanics is a work in progress. Adjustments to the theory and reinterpretations are proposed to this day. Though textbooks may give the opposite impression, not everyone was completely pleased with the Copenhagen interpretation. It may be recalled that Einstein was never wholly comfortable with quantum mechanics. Louis de Broglie, likewise, kept revisiting the system for other possible explanations or interpretations.

But other proposed theories have, for various reasons, fallen by the wayside. One such theory was proposed by David Bohm (1917–1992), and he defended it his entire life. David Bohm expressed fears that the "bigshots" would handle his theory with a "conspiracy of silence" so his theories, from a "smaller shot," would seem unimportant.[11] This is basically what happened.

David Bohm did not start out with a rough life, but it got rougher. Born in Pennsylvania, he did undergraduate work in physics at Pennsylvania State College and graduate work at Berkeley under Oppenheimer. He became assistant professor at Princeton University in 1946 and then in 1949 was cited for contempt of Congress by the House Committee on Un-American Activities. Oppenheimer, himself persecuted, began turning in to the Federal Bureau of Investigation names of friends and acquaintances who he thought might be communist agents. Bohm apparently was one of the accused.[12] In 1951 Bohm was forced to leave Princeton University and was unable to find employment as a physicist anywhere else in the United States. The next year, when he published his alternate formulation of quantum mechanics, it met with a cool reception. True, the theory made no initial predictions that were different from the Copenhagen formulation, but then no Dirac rose up to investigate the relativistic implications and no Pauling applied it to chemical bonding. Bohm's theory was basically ignored or passed over as superfluous.

In Bohm's theory particles are particles and waves are waves. There is no wave–particle duality. The particles are moved around by fields, but not the types of fields we have dealt with before, such as electric or magnetic fields. Electric and magnetic fields die off with distance and they exert a force on the particles with which they interact. Bohm's fields are called pilot waves because they pilot the particles around, but do not change the energy of the particles and do not die off with distance. The pilot waves do provide an explanation for entanglement: When one particle is spin-up, it communicates this information to its symmetric partner via the pilot wave that contains them both. The advantage to Bohm's theory is that wave–particle duality is gone, but something else is back that we thought we'd gotten rid of: The pilot wave sounds suspiciously like ether. So it was a name-your-poison situation. Granted, the notion of mystical undying fields that pervade everything and change nothing is hard to swallow, but so are cats killing other cats.

Bohm wandered to Brazil, then Israel, then England, where he finally settled in 1961. He continued to work on his theory, though he eventually turned to other pursuits. He discontinued his interest in Marxism, if indeed he ever had much of any, and became a follower of the Indian guru Jiddo Krishnamurti.[13] After many years of association with the guru, Bohm lost faith when it was revealed that the spiritual leader was also an adulterer. Bohm had received psychotherapy for depression, and now he needed to be hospitalized. He underwent shock therapy for his mental state and heart surgery, later dying of a massive heart attack at the age of seventy-five.

Did Bohm have a "secret theory" that held a "truth" and has been "suppressed"? Probably not. Does his work demonstrate that there are other ways to model atomic level physics? Yes.

Other approaches to solving the mystery of superposition and entanglement have included a search for some hidden variables that might somehow be controlling the outcome of the experiments. In fact, it was this search for hidden variables that prompted John Bell (1928–1990) to suggest the set of experiments on spin we described above. These experiments, if they are valid, show no uncontrolled variables have been detected so far. An interesting "many worlds" explanation was put forward by Hugh Everett III in his Princeton doctoral thesis.[14] In this equally difficult-to-fathom interpretation, each of the outcomes is equally real, and indeed the observer is in two different mental states and so able to perceive fully the two different outcomes. The advantage is mathematical: It simplifies the treatment (if such a thing is possible) but requires mental gymnastics to ingest.[15] Another approach, proposed by Dieter Zeh of the University of Heidelberg, is to say that *decoherence*, that is, the condition of being in several states at once, is really the nature of the world. The reason we don't perceive the world as such is because any contact

Fig. 34. David Bohm. Photograph taken May 1949 after Bohm refused to testify before the House Un-American Activities. ACME Telephoto, courtesy AIP Emilio Segrè Visual Archives.

with the environment suffices to collapse the superposition of states. So being struck with an air particle or light ray is enough to collapse this dual reality to the single reality that we are used to.

But is all of this creativity and burning of mental energy necessary? We have said that quantum mechanics and quantum chemistry are already used in virtually all branches of chemistry and atomic and chemical physics. Why is there need for more speculation? Because if a new formulation of quantum mechanics could be found, it might bring us closer to a major goal in theoretical science. In the context of nanotechnology we mentioned that Richard Feynman had received his Nobel Prize for quantum electrodynamics (shared with Julian Schwinger and Shinichiro Tomonaga), but we didn't mention what quantum electrodynamics is. Quantum electrodynamics is a more complete description of the interaction between light and matter, and calculations based on it have been shown to be correct to better than one part in a thousand. In addition, the theory combines Maxwell's electromagnetic wave with quantum mechanics and special relativity to give a united theory. This discovery, in turn, has encouraged scientists to believe that there may exist one grand "theory of everything"[16] that will unite the remaining disparate theories—quantum mechanics, internuclear forces, and general relativity—under one Grand Unified Theory, or GUT,[17] the current Holy Grail of science.

But what do internuclear forces have to do with physical chemistry? Isn't that nuclear physics? Doesn't the concern of the physical chemist stop at quantum mechanics?

I believe we have time for one more story.

EYNDE

FINIS CORONAT.

Epilogue

I think we have no cage for such a bird.

S. H. Burbury, circa 1900

Marie Mayer (1906–1972) started life as Maria Gertrud Käte Geoppert, the only child of a German professor of pediatrics and his wife.[1] Göttingen, the town in which she was born, has stone walls dating to the German middle ages, and Marie and her father would take walks together in the forest.

Marie's parents encouraged her education. She started out studying mathematics, but then became acquainted with quantum mechanics and decided to study physics as well. Max Born, her mentor, spoke of her in glowing terms. Her doctoral dissertation consisted of an extension of Dirac's work on the theory of the interaction of light with matter.

Joseph Mayer was a young American chemist who had recently obtained his degree while working with Gilbert Lewis at Berkeley. In wandering to the shrines of science in Europe, he met Marie when he was a boarder in her home. Marie and Joe married in Germany while Marie was completing her degree, then moved to the United States so Joe could assume an assistant professorship in chemistry at Johns Hopkins.

Most large universities employ a considerable part-time instructional staff to lighten the teaching load of their full-time research professors. Following a well-precedented pattern, Marie was employed at the university in this capacity. As her husband was part of the permanent staff, she was given a desk and work space, though much of her work was without compensation. After a few years at Johns Hopkins, she had basically assumed the role of an assistant professor—teaching, doing research, and taking on students—but without a formal appointment. This situation did not seem to pose any particular difficulty for her. There were nepotism rules that did not allow her to be officially employed while her husband was on the faculty. She was also used to the European system that could keep an academic working strictly on students' fees for an entire career. She was busy and she was happy.

Adapting her interests to her surroundings, she became a physical chemist. She taught mostly chemistry and collaborated in theoretical chemical research. She wrote a textbook with her husband on the application of statistical mechanics to chemistry that went through many printings.

During this period Marie and Joe also became parents to a son and a daughter. Domestic demands began to occupy her, including hosting faculty gatherings in their home. It was the 1930s and the jazz age. During Prohibition the Mayers practiced their chemistry making wine, and they were said to be the last to leave the dance floor at many a late night party. Then the Depression, perhaps combined with some xenophobic reactions to Joe's German wife, resulted in Joe's being dismissed from Johns Hopkins. Harold Urey offered Joe a position at Columbia University, so the family moved to New York. Marie again had no official appointment, but she routinely gave lectures and was able to continue her research.

With the advent of World War II, Urey was asked to assemble a group to work on the problem of isotope separation for the Manhattan Project, and he recruited Marie. It may be recalled that we first met Marie in this context, investigating isotope effects on reaction rates. The project required her to use her knowledge of symmetry, group theory, quantum mechanics, statistical mechanics, thermodynamics, and spectroscopy to solve a problem in radiochemistry, and Marie thrived on this work. Joe was called to the Aberdeen Proving Grounds in Maryland to work on conventional weapons and was home only one day a week. This partial separation seems to have been

Fig. 35. Marie Goeppert Mayer and Enrico Fermi, circa 1930. AIP Emilio Segrè Visual Archives, Goudsmit Collection.

pivotal for Marie: She was now required to work independently and was responsible for the work of a group of professionals under her supervision.

Despite the stimulating research, this was a difficult time for Marie. She still had friends and family in Germany, she was isolated from her husband, and she had less time to spend with her children. Her work, however, remained penetrating, quick, and incisive.[2] After the war, Joe and Marie accepted positions at the University of Chicago, where she now held an official appointment as associate professor, but because of nepotism rules it still had to be voluntary, an arrangement that was not unusual at the time for married academics. She was also offered a half-time position as a research physicist at the Argonne National Lab, another turning point in her life.

She did not know much about nuclear physics at the time but was used to the idea of learning a new field when the job required it. As it turned out, her lack of knowledge was to her advantage. In 1947, sorting through some data regarding isotopes, Marie saw that certain numbers of neutrons and protons kept showing up more often than others, and she became convinced there was a pattern. After a mathematician derisively called her numbers "magic numbers," she adopted the phrase.

She published her findings, only to find that others before her had noted the same coincidence, although she had made the relationship stronger with additional evidence. Yet there was no theory to go with the data. So Marie kept playing with various models, hoping to find an explanation for the magic numbers. When Marie found that spin could be assigned to the nucleus, and that it varied depending on the number of nucleons that the nucleus contained (that is, protons and neutrons), she had an inkling that this property might somehow be connected. She discussed the idea with Enrico Fermi, the Nobel laureate she had become acquainted with through her university teaching and work on the Manhattan Project.

As the story is told, Fermi mentioned that the coupling between electron spin and orbital angular momentum contributed to the observed shell structure of atoms—at which point he was called to the telephone—and by the time he returned from his phone call, Marie had fit the idea of spin coupling into her nuclear shell model.[3] Spin-orbit coupling might be understood by considering how a baseball thrown with a spin results in a curve ball (though the actual mechanism for a curve ball involves air resistance).[4] A ball will curve left if thrown with a counterclockwise spin, and it will curve right if thrown with a clockwise spin. Thus the two trajectories are "split" into left and right. When Marie incorporated the coupling of the spins of the nucleons in her model, she found energy level splittings at precisely her magic numbers.

Why had this idea been missed by others? Because the leading authorities of the day had decided the forces within the nucleus precluded independent spins for the nucleons; the protons and neutrons were just too tightly bound. The dominant model of the day, the liquid-drop model, which essentially treats the nucleus as one entity instead of composed of individual particles, had been used to explain the fission reactions of the Manhattan Project. Nobody seemed interested in modifying it. But Marie hadn't been raised in the faith and she wasn't worried about losing her reputation because she had no reputation to lose. She used her knowledge of physics, but as her biographer notes, she approached the problem as a chemist, patiently compiling large amounts of data and sifting through to find a pattern.[5] She was a mathematician turned physicist turned chemist, and this was her great strength: She had the ability to assume all roles and cross all boundaries, because for her boundaries did not exist.

In 1948, at the age of forty-two, she published her theory in a short letter, so brief that is was almost overlooked, but so powerful in its simplicity that it soon gained recognition. Experimental evidence was found to support the predictions of her theory, and in 1959 the University of California at San Diego offered both Joe and Marie full, paid professorships. At this point the University of Chicago decided it was time to drop the nepotism rule. But the Mayers opted for California. Unfortunately, within a month of arriving, Marie suffered a stroke. Although she recovered, her left arm was essentially paralyzed and she had difficulty speaking. She continued her work and travels, but she never completely recovered. When awarded the Nobel Prize in 1963, at the age of fifty-seven, she was concerned that she would not be able to navigate through the ceremony given her physical condition. When the time came, however, she managed. She accepted the gold medal and the diploma and offered her hand to the king. Turned to the king as she was, she could not see that her husband was crying.[6]

Breakthroughs come from those who can see connections where no one else can. Those who see invisible light. Those who throw a spear where the fish isn't. Are breakthroughs still out there to be made? Yes. Friction and intermolecular forces have yet to be fully elucidated. Energy and entropy cannot be calculated instantly for any arbitrary reaction. The rates of reactions cannot be predicted neatly from first principles. The mechanisms for many complicated reactions, including biologically important interactions, have yet to be fully mapped out. The ultimate possibilities of nanotechnology await theoretical as well as experimental development. Chemical oscillations and waves, those strange, symmetric structures that spring from seemingly chaotic solutions, have yet to be completely understood. And all that quantum weirdness . . .

But progress will come. Progress may come in the form of a new fundamental principle. Like Newton's gravity. Or Boyle's gas law. Or Dalton's atoms. Or Gibbs's free energy. Or Maxwell's electromagnetic wave. Or Faraday's fields. Or Noether's symmetry. Or Clausius's entropy. Or Lewis's two-electron bond.

Progress may come when someone with a different perspective jumps into the fray. An iconoclast like Lewis. Or accountant like Lavoisier. Or bohemian like Schrödinger. Or gambler like Cardano. Or eclectic like Aristotle. Or perfectionist like Perrin. Or recluse like Gibbs. Or dissenter like Dalton. Or magician like della Porta. Or hausfrau like Mayer.

Progress will come when the time is right—though it is impossible to say whose services it will choose. Progress will be made by someone with imagination, vision, and valor, too. Some rebel, some radical, some maverick. Some spirit, who for some reason, insists on being free.

Notes

Introduction

1. Anthony Standen, *Science Is a Sacred Cow* (New York: Dutton, 1950), pp. 77–78.

Part I. Aristotle and the Ancients

Introduction

1. Alan and Veronica Palmer, comp. *Quotations in History: A Dictionary of Historical Quotations c. 800 A.D. to the Present* (New York: Barnes & Noble Books, 1976), pp. 48–129.
2. J. D. Bernal, *History of Classical Physics, from Antiquity to the Quantum* (New York: Barnes & Noble Books, 1972), p. 104.

Chapter 1. The Ancients

1. J. Hacker, ed., *Grolier Encyclopedia of Knowledge*, 20 vols. (Danbury, Conn.: Grolier, 1993), 9: 21.
2. Richard Boehm et al., *The World: Harcourt Brace Social Studies* (New York: Harcourt Brace, 2000), p. 229.
3. Theophile Obenga, *Ancient Egypt and Black Africa, A Student's*

Handbook for the Study of Ancient Egypt in Philosophy, Linguistics, and Gender Relations, trans. Sylvianne Martinon and Ahmed Sheik (London: Karnak House, 1992), p. 51.

4. C. C. Gillispie, ed., *Dictionary of Scientific Biography*, 18 vols. (New York: Scribner, 1970–1990), 13: 295–98.

5. D. Bergamini, ed., *Mathematics* (New York: Time, 1963), pp. 156–58.

6. Gillispie, *Dictionary of Scientific Biography*, 4: 414–59.

7. Ibid., 15: 207–24 (Hipparchus), 11: 186–206 (Ptolemy).

8. Ibid., 11: 22–31 (Plato).

Chapter 2. Aristotle

1. Dava Sobel, *Galileo's Daughter: A Historical Memoir of Science, Faith, and Love* (New York: Walker & Co., 1999), p. 53.

2. C. C. Gillispie, ed., *Dictionary of Scientific Biography*, 18 vols. (New York: Scribner, 1970–1990), 1: 250–81.

3. J. D. Bernal, *History of Classical Physics, from Antiquity to the Quantum* (New York: Barnes & Noble Books, 1972), p. 92.

4. *The Works of Aristotle*, W. Benton, ed., 2 vols. (Chicago: Encyclopedia Britannica, 1952), 1: v.

5. Eleanor J. Gibson, *An Odyssey in Learning and Perception* (Cambridge, Mass.: MIT Press, 1991), pp. 103–105.

6. Quoted in Gillispie, *Dictionary of Scientific Biography*, 1: 250.

7. Eric Haseltine, "The Greatest Unanswered Questions of Physics," *Discover* (February 2002): 37.

8. Gillispie, *Dictionary of Scientific Biography*, 4: 465–67.

9. Ibid., 2: 387–89.

10. Ibid., 4: 465–67.

11. William Osler, *The Evolution of Modern Medicine* (New Haven, Conn.: Yale University Press, 1921), p. 82.

Chapter 3. The Arabs

1. W. Montgomery Watt, *The Influence of Islam on Medieval Europe* (Edinburgh: University Press, 1972), p. 7.

2. Victor Katz, *A History of Mathematics* (New York: HarperCollins, 1992), p. 228.

3. C. C. Gillispie, ed., *Dictionary of Scientific Biography*, 18 vols. (New York: Scribner, 1970–1990), 7: 323–34.

4. Ibid., 7: 358–65.

5. Ibid., 7: 335–51.

6. Richard Boehm et al., *The World: Harcourt Brace Social Studies* (New York: Harcourt Brace, 2000), p. 363.

7. Watt, *The Influence of Islam on Medieval Europe*, pp. 65–66.

8. Marco Polo, *The Travels of Marco Polo*, trans. Ronald Latham (New York: Penguin Books, 1982), p. 156. See also John Larner, *Marco Polo and the Discovery of the World* (New Haven, Conn.: Yale University Press, 1999).

Chapter 4. Medieval Magick and Renaissance Revival

1. J. D. Bernal, *History of Classical Physics, from Antiquity to the Quantum* (New York: Barnes & Noble Books, 1972), pp. 124–26.

2. John Mann, *Murder, Magic, and Medicine* (New York: Oxford University Press, 1992), pp. 76–82.

3. Barbara Ehrenreich, *For Her Own Good: 150 Years of the Experts' Advice to Women* (Garden City, N.Y.: Anchor Books, 1979), p. 35.

4. Ibid., p. 37.

5. H. R. Trevor-Roper, *The European Witch-Craze of the Sixteenth and Seventeenth Centuries, and Other Essays* (New York: Harper & Row, 1969), pp. 90–192.

6. C. C. Gillispie, ed., *Dictionary of Scientific Biography*, 18 vols. (New York: Scribner, 1970–1990), 1: 196–200.

7. J. Hacker, ed., *Grolier Encyclopedia of Knowledge*, 20 vols. (Danbury, Conn.: Grolier, 1993), 15: 229.

8. Gillispie, *Dictionary of Scientific Biography*, 11: 95; Louise George Clubb, *Giambattista Della Porta* (Princeton, N.J.: Princeton University Press, 1965).

9. John Baptista Porta, *Natural Magick* (London: Printed for T. Young and S. Speed, 1658). May also be found at http://members.tscnet.com/pages/omard1/jportat2.html.

10. Jacob Bronowski, *Magic, Science, and Civilization* (New York: Columbia University Press, 1978), p. 31.

11. Lynn Thorndike, *A History of Magic and Experimental Science*, 8 vols. (New York: Columbia University Press, c. 1923–1958) 7: 227, 238, 339, 380, 660; 8: 34, 196, 202, 553, 606, 633; Penelope Gouk, *Music, Science, and Natural Magic in Seventeenth-Century England* (New Haven, Conn.: Yale University Press, 1999).

12. Porta, *Natural Magick*, 9: chap. 29.

Chapter 5. The Merchants

1. C. C. Gillispie, ed., *Dictionary of Scientific Biography*, 18 vols. (New York: Scribner, 1970–1990), 4: 604–12.
2. D. Bergamini, ed., *Mathematics* (New York: Time, 1963), p. 93.
3. Ibid., p. 67.
4. Gillispie, *Dictionary of Scientific Biography*, 8: 204.
5. Diane Stanley, *Leonardo Da Vinci* (New York: Morrow, 1996), p. 10.
6. Ibid., p. 16.
7. Gillispie, *Dictionary of Scientific Biography*, 4: 258–61.
8. Bert S. Hall, *Weapons and Warfare in Renaissance Europe: Gunpowder, Technology, and Tactics* (Baltimore: Johns Hopkins University Press, 1997), p. 192.

Part II. The European Scientific Revolution

Introduction

1. *The Oxford English Dictionary*, 2d ed., 20 vols. (Oxford: Clarendon Press, 1989), 11: 747.
2. Ibid., 2: 84.
3. H. R. Trevor-Roper, *The European Witch-Craze of the Sixteenth and Seventeenth Centuries, and Other Essays* (New York: Harper & Row, 1969), p. 169.
4. C. C. Gillispie, ed., *Dictionary of Scientific Biography*, 18 vols. (New York: Scribner, 1970–1990), 13: 573.
5. Bert S. Hall, *Weapons and Warfare in Renaissance Europe, Gunpowder, Technology, and Tactics* (London: Johns Hopkins University Press, 1997), p. 20.
6. John Baptista Porta, *Natural Magick* (London: Printed for T. Young and S. Speed, 1658). May also be found at http://members.tscnet.com/pages/omard1/jportat2.html.
7. Gillispie, *Dictionary of Scientific Biography*, 6: 130.

Chapter 6. Mathematics

1. C. C. Gillispie, ed., *Dictionary of Scientific Biography*, 18 vols. (New York: Scribner, 1970–1990), 3: 64–67 (Cardano), 13: 258–62 (Tartaglia).

2. D. Bergamini, ed., *Mathematics* (New York: Time, 1963), p. 68.

3. Gillispie, *Dictionary of Scientific Biography*, 4: 595–96.

4. Victor J. Katz, *A History of Mathematics* (New York: Harper-Collins, 1993), p. 330.

5. Bergamini, *Mathematics*, p. 69.

6. Girolamo Cardano, *The Great Art,* trans. and ed. T. Richard Witmer (Cambridge, Mass.: M.I.T. Press, 1968), pp. 98–99.

7. Katz, *A History of Mathematics*, p. 324.

8. Gillispie, *Dictionary of Scientific Biography*, 14: 18–25.

9. Katz, *A History of Mathematics*, p. 340.

10. Gillispie, *Dictionary of Scientific Biography*, 7: 47–51.

11. Theoni Pappas, *Mathematical Scandals* (San Carlos, Calif.: Wide World Publishing, 1997) pp. 1–7.

12. Gillispie, *Dictionary of Scientific Biography*, 9: 611.

13. Pappas, *Mathematical Scandals*, p. 109.

14. Lynn Thorndike, *A History of Magic and Experimental Science* (New York: Columbia University Press, 1958), p. 188.

15. Gillispie, *Dictionary of Scientific Biography*, 13: 573–74.

Chapter 7. Physics

1. C. C. Gillispie, ed., *Dictionary of Scientific Biography*, 18 vols. (New York: Scribner, 1970–1990), 5: 238.

2. *The Mechanical Universe—and Beyond*, videorecording, The Annenberg/CPB collection (Santa Barbara, Calif.: Intellimation [distributor], 1987), videocassette 1, segment 1.

3. Gillispie, *Dictionary of Scientific Biography*, 5: 237–49.

4. Dava Sobel, *Galileo's Daughter: A Historical Memoir of Science, Faith, and Love* (New York: Walker & Co., 1999), pp. 19–20.

5. Gillispie, *Dictionary of Scientific Biography*, 5: 237–49.

6. Ibid.

7. Sobel, *Galileo's Daughter*, p. 19.

8. Gillispie, *Dictionary of Scientific Biography*, 3: 401–11.

9. Ibid., 2: 539–44.

10. Sobel, *Galileo's Daughter*, pp. 156–57.

11. Ibid., pp. 71–83.

12. Ibid.

Chapter 8. Chemistry

1. Dava Sobel, *Galileo's Daughter: A Historical Memoir of Science, Faith, and Love* (New York: Walker & Co., 1999), p. 200.

2. Lloyd O. Bishop and Will S. Deloach, "Marie Meurdrac—First Lady of Chemistry?" *Journal of Chemical Education* 47 (1970): 448–49.

3. Marelene Rayner-Canham and Geoffrey Rayner-Canham, *Women in Chemistry: Their Changing Roles from Alchemical Times to the Mid-Twentieth Century* (New York: American Chemical Society, 1998), pp. 6–7.

4. Marie Meurdrac, *La Chymie Charitable and Facile, en Faveur Des Dames* (Paris: CNRS Editions, 1999).

5. Cathy Cobb and Harold Goldwhite, *Creations of Fire* (New York: Plenum, 1995).

6. C. C. Gillispie, ed., *Dictionary of Scientific Biography*, 18 vols. (New York: Scribner, 1970–1990), 6: 254.

7. James Riddick Partington, *A History of Chemistry*, 4 vols. (New York: St. Martin's Press, 1961), 2: 15, 30, 83, 177–78.

8. Ibid., 2: 486–49.

9. Ibid., 2: 525.

Chapter 9. Mathematics and Physics

1. Lloyd Motz and Jefferson Hane Weaver, *Story of Physics* (New York: Plenum, 1989), p. 49.

2. C. C. Gillispie, ed., *Dictionary of Scientific Biography*, 18 vols. (New York: Scribner, 1970–1990), 10: 42–103.

3. Victor J. Katz, *A History of Mathematics* (New York: Harper-Collins, 1993), p. 404.

4. Jennifer Lee Carrell, "Newton's Vice," *Smithsonian* (December 2000): 131–46.

5. Ibid.

6. Katz, *A History of Mathematics*, p. 477.

Chapter 10. Lavoisier

1. Marelene Rayner-Canham and Geoffrey Rayner-Canham, *Women in Chemistry: Their Changing Roles from Alchemical Times to the Mid-Twentieth Century* (Washington, D.C.: American Chemical Society, 1998), p. 15.

2. Louise S. Grinstein et al., ed., *Women in Chemistry and Physics: A*

Bibliographic Sourcebook (Westport, Conn.: Greenwood Press, 1993), pp. 101–105.

3. Jean-Pierre Poirier, *Lavoisier, Chemist, Biologist, Economist*, trans. Rebecca Balinski (Philadelphia: University of Pennsylvania Press, 1996).

4. Ibid.

5. Ibid., p. 23.

6. Ibid.

7. James Riddick Partington, *A History of Chemistry*, 4 vols. (New York: St. Martin's Press, 1961), 3: 368.

8. Frederic L. Holmes, "Antoine Lovoisier and the Conservation of Matter," *Chemical and Engineering News* (September 12, 1994): 38–45.

9. C. C. Gillispie, ed., *Dictionary of Scientific Biography*, 18 vols. (New York: Scribner, 1970–1990), 8: 66–91.

10. Partington, *A History of Chemistry*, 3: 36.

11. Gillispie, *Dictionary of Scientific Biography*, 8: 82.

12. Grinstein et al., *Women in Chemistry and Physics*, pp. 314–19.

13. Poirier, *Lavoisier*, p. 394.

14. Ibid., p. 409.

Part III. The First Atomic Wars

Introduction

1. Keith J. Laidler, *The World of Physical Chemistry* (New York: Oxford University Press, 1995), pp. 46–52.

2. Rene J. Dubos, *Louis Pasteur, Free Lance of Science* (Boston: Little, Brown, 1950), p. 39.

3. Margaret W. Rossiter, "Mendel the Mentor," *Journal of Chemical Education* 71 (1994): 215–19.

4. Daniel Fallon, *The German University: A Heroic Ideal in Conflict with the Modern World* (Boulder: Colorado Associated University Press, 1980).

5. Mary Jo Nye, *Before Big Science, the Pursuit of Modern Chemistry and Physics, 1800–1940* (Cambridge, Mass.: Harvard University Press, 1996), p. 29.

Chapter 11. Dalton's Diminutive Friends

1. C. C. Gillispie, ed., *Dictionary of Scientific Biography*, 18 vols. (New York: Scribner, 1970–1990), 3: 537–47.

2. Ibid., 2: 377–82.

3. *The Oxford English Dictionary*, 2d ed., 20 vols. (Oxford: Clarendon Press, 1989), 16: 746.

4. William Hodson Brock, *The Norton History of Chemistry* (New York: W. W. Norton, 1993), p. 132.

5. Mary Jo Nye, *Before Big Science, the Pursuit of Modern Chemistry and Physics, 1800–1940* (Cambridge, Mass.: Harvard University Press, 1996), pp. 13–15, 29.

6. Ibid., p. 36.

7. Gillispie, *Dictionary of Scientific Biography*, 6: 284–85.

8. Ibid., 4: 238–42.

9. James Riddick Partington, *A History of Chemistry*, 4 vols. (New York: St. Martin's Press, 1961), 4: 199.

10. Gillispie, *Dictionary of Scientific Biography*, 10: 545–46.

11. Karen Timberlake, *Chemistry and Life* (New York: Addison Wesley, 2001), p. 403.

12. Stephen G. Brush, *The Kind of Motion We Call Heat: A History of the Kinetic Theory of Gases in the 19th Century*, Book 1, *Physics and the Atomists* (Amsterdam: North-Holland Publishing, 1976), p. 295.

13. Brock, *Norton History of Chemistry*, p. 134.

14. Ibid., p. 158; Gillispie, *Dictionary of Scientific Biography*, 2: 90–96.

Chapter 12. Thermodynamics

1. C. C. Gillispie, ed., *Dictionary of Scientific Biography*, 18 vols. (New York: Scribner, 1970–1990), 14: 184–86.

2. *The New Encyclopedia Britannica*, 12 vols. (Chicago: Encyclopedia Britannica, 1998), 11: 229.

3. Gillispie, *Dictionary of Scientific Biography*, 10: 36–37.

4. Keith J. Laidler, *To Light Such a Candle* (Oxford: Oxford University Press, 1998), p. 19.

5. Gillispie, *Dictionary of Scientific Biography*, 11: 350.

6. Stephen G. Brush, *The Kind of Motion We Call Heat: A History of the Kinetic Theory of Gases of the 19th Century*, Book 2, *Statistical Physics and Irreversible Processes* (New York: North-Holland Publishing, 1976), p. 524.

7. Gillispie, *Dictionary of Scientific Biography*, 7: 180–82.

8. Ibid., 13: 374–88.

9. Ibid., 1: 138.

10. Ibid., 6: 241–53.

11. Ibid., 3: 79–84.

12. Ibid., 3: 303–10.

13. This analogy was borrowed, with thanks, from Thomas Colbert, Augusta State University.

Chapter 13. Gnats in Sunbeams

1. C. C. Gillispie, ed., *Dictionary of Scientific Biography*, 18 vols. (New York: Scribner, 1970–1990), 2: 36–56.

2. Ibid., 2: 43.

3. Ibid., 6: 293–94.

4. Ibid., 5: 492–95.

5. Ibid., 14: 184–86.

6. Stephen G. Brush, *The Kind of Motion We Call Heat: A History of the Kinetic Theory of Gases in the 19th Century*, Book 1, *Physics and the Atomists* (Amsterdam: North-Holland Publishing, 1976), p. 136.

7. Ibid.

8. Gillispie, *Dictionary of Scientific Biography*, 14: 185.

9. Brush, *The Kind of Motion We Call Heat*, p. 136.

10. Ibid., p. 143.

11. Ibid., p. 246.

Chapter 14. Statistical Mechanics

1. Girolamo Cardano, *The Book on Games of Chance* (New York: Holt, Rinehart, Winston 1953), pp. 32–35.

2. Ibid., p. 11.

3. D. Bergamini, ed., *Mathematics* (New York: Time, 1963), p. 128.

4. Keith J. Laidler, *The World of Physical Chemistry* (New York: Oxford University Press, 1995), p. 65.

5. C. C. Gillispie, ed., *Dictionary of Scientific Biography*, 18 vols. (New York: Scribner, 1970–1990), 6: 323–28.

6. Ibid., 12: 524.

7. Ibid., 9: 198–230.

8. Ibid., 9: 228, n. 71.

9. Keith J. Laidler, *To Light Such a Candle* (Oxford: Oxford University Press, 1998), p. 48.

10. Gillispie, *Dictionary of Scientific Biography*, 2: 260–68.

11. Stephen G. Brush, *The Kind of Motion We Call Heat: A History of the Kinetic Theory of Gases in the 19th Century*, Book 2, *Physics and the Atomists* (Amsterdam: North-Holland Publishing, 1976), p. 602.

12. Paul G. Hewitt, *Conceptual Physics*, 11th ed. (New York: Addison-Wesley, 1998), p. 104.

13. Gillispie, *Dictionary of Scientific Biography*, 13: 139.

14. Ibid., 7: 526, 14: 170–72.

15. Ibid., 9: 223.

Chapter 15. Thermochemistry

1. Karen Timberlake, *Chemistry*, 7th ed. (New York: Addison Wesley Longman, 1999), p. 273.

2. Keith J. Laidler, *To Light Such a Candle* (Oxford: Oxford University Press, 1998), p. 49.

3. C. C. Gillispie, ed., *Dictionary of Scientific Biography*, 18 vols. (New York: Scribner, 1970–1990), 5: 286–393; Muriel Rukeyser, *Williard Gibbs* (Woodbridge, Conn.: Ox Bow Press, 1942), pp. 94–95.

4. Ibid., chap. 2.

5. Clifford E. Dykstra, *Physical Chemistry—A Modern Introduction* (Upper Saddle River, N.J.: Prentice-Hall, 1997), p. 134.

6. William Hodson Brock, *The Norton History of Chemistry* (New York: W. W. Norton, 1993), pp. 644–47.

Chapter 16. Atoms or Not?

1. C. C. Gillispie, ed., *Dictionary of Scientific Biography*, 18 vols. (New York: Scribner, 1970–1990), 8: 507–11; James Riddick Partington, *A History of Chemistry*, 4 vols. (New York: St. Martin's Press, 1961), 4: 546.

2. Keith J. Laidler, *The World of Physical Chemistry* (New York: Oxford University Press, 1995), p. 348.

3. *The Oxford English Dictionary*, 2d ed., 20 vols. (Oxford: Clarendon Press), 9: 971.

4. Gillispie, *Dictionary of Scientific Biography*, 15: 455–69.

5. Paul G. Hewit et al., *Conceptual Physical Science* (New York: HarperCollins, 1994), p. 466.

6. Nicholas Tzimopoulos, *Modern Chemistry* (New York: Holt, Rinehart and Winston, 1990), p. 88.

7. Kenneth Witten et al., *General Chemistry,* 6th ed. (New York: Saunders, 2000), p. 61.

8. Mary Jo Nye, *Molecular Reality: A Perspective on the Scientific Work of Jean Perrin* (New York: American Elsevier, 1972).

9. Gillispie, *Dictionary of Scientific Biography,* 2: 516–22.

10. Stephen G. Brush, *The Kind of Motion We Call Heat: A History of the Kinetic Theory of Gases in the 19th Century,* Book 2, *Physics and the Atomists* (Amsterdam: North-Holland Publishing, 1976), p. 657.

11. Ibid.

12. Gillispie, *Dictionary of Scientific Biography,* 10: 524–26.

13. Nye, *Molecular Reality.*

14. Ibid., p. 135.

15. Gillispie, *Dictionary of Scientific Biography,* 2: 464.

16. Brush, *The Kind of Motion We Call Heat,* p. 295.

Part IV. Physics and Chemistry Come to Light

Introduction

1. C. C. Gillispie, ed., *Dictionary of Scientific Biography,* 18 vols. (New York: Scribner, 1970–1990), 13: 295–98. See also Jennifer Duellette, "Teaching Old Masters New Tricks," *Discover* (December 2001): 82–83.

2. Gillispie, *Dictionary of Scientific Biography,* 12: 79.

3. Keith J. Laidler, *The World of Physical Chemistry* (New York: Oxford University Press, 1995), p. 250.

4. Keith J. Laidler, *To Light Such a Candle* (Oxford: Oxford University Press, 1998), p. 71.

5. Gillispie, *Dictionary of Scientific Biography,* 8: 388.

Chapter 17. Spectroscopy

1. C. C. Gillispie, ed., *Dictionary of Scientific Biography,* 18 vols. (New York: Scribner, 1970–1990), 10: 79.

2. Ibid., 6: 328–36.

3. Ibid., 11: 473–75.

4. Ibid., 11: 473.

5. Ibid., 5: 538.

6. Ibid., 5: 142.

7. Ibid., 13: 237–39 (Talbot); 6: 237–39 (Herschel).

8. Keith J. Laidler, *The World of Physical Chemistry* (New York: Oxford University Press, 1995), p. 167.

9. Gillispie, *Dictionary of Scientific Biography*, 2: 45.

10. Peter Atkins, *Physical Chemistry*, 5th ed. (New York: Freeman, 1994), p. 602.

11. Gillispie, *Dictionary of Scientific Biography*, 7: 380 (Kirchnoff), 2: 586 (Bunsen).

12. Nicholas C. Thomas, "The Early History of Spectroscopy," *Journal of Chemical Education* 66 (1991): 631.

13. Laidler, *The World of Physical Chemistry*, p. 168.

14. Gillispie, *Dictionary of Scientific Biography*, 2: 586.

15. Ibid., 8: 440–43.

16. Quoted in Ibid., 8: 440.

17. Ibid., 7: 186.

18. Michael Epstein, "Lecoq De Boisbadran, the Unknown Spectroscopist," *The Spectrum*, 27 (September 2000): 2; Lloyd O. Bishop and Will S. Deloach, "Marie Meurdrac—First Lady of Chemistry?" *Journal of Chemical Education* 47 (1970): 448–49.

19. Gillispie, *Dictionary of Scientific Biography*, 1: 425.

Chapter 18. Electromagnetism

1. C. C. Gillispie, ed., *Dictionary of Scientific Biography*, 18 vols. (New York: Scribner, 1970–1990), 14: 562.

2. Stephen G. Brush, *The Kind of Motion We Call Heat: A History of the Kinetic Theory of Gases in the 19th Century*, Book 1, *Physics and the Atomists* (Amsterdam: North-Holland Publishing, 1976), p. 42.

3. Gillispie, *Dictionary of Scientific Biography*, 15: 480–90.

4. Alfred Still, *Soul of Amber: The Background of Electrical Science* (New York: Murray Hill Books, 1944), p. 183.

5. Gillispie, *Dictionary of Scientific Biography*, 4: 527–40.

6. L. Pearce Williams, *Michael Faraday* (New York: Da Capo, 1987), pp. 18–19.

7. Still, *Soul of Amber*, pp. 209–19.

8. Seymour Simon, *Sharks* (New York: Scholastic, 1995).

9. Cited in Gillispie, *Dictionary of Scientific Biography*, 9: 206.

10. Richard Wolfson, *Einstein's Relativity and the Quantum Revolution: Modern Physics for Non-Scientists,* part 1 (Springfield, Va.: The Teaching Co., 1999), tape 4.

11. Hugh Hildreth Skilling, *Exploring Electricity; Man's Unfinished Quest* (New York: Ronald Press Co., 1948), p. 174.

12. Gillispie, *Dictionary of Scientific Biography,* 6: 340.

Chapter 19. Atomic Structure

1. L. Pearce Williams, *Michael Faraday* (New York: Da Capo, 1987), p. 242.

2. C. C. Gillispie, ed., *Dictionary of Scientific Biography,* 18 vols. (New York: Scribner, 1970–1990), 13: 82.

3. Ibid., 5: 458.

4. Ibid., 13: 362–72.

5. Keith J. Laidler, *To Light Such a Candle* (Oxford: Oxford University Press, 1998), p. 242.

6. Gillispie, *Dictionary of Scientific Biography,* 11: 529–31.

7. Ibid., 1: 558–61.

8. Ibid., 3: 474–80.

9. Arthur E. Woodruff, "The Radiometer and How It Does Not Work," *The Physics Teacher* (October 1968): 358–63.

10. Gillispie, *Dictionary of Scientific Biography,* 9: 542–45.

11. Linda Merricks, *The World Made New: Frederick Soddy, Science, Politics, and Environment* (New York: Oxford University Press, 1996), p. 70.

12. Gillispie, *Dictionary of Scientific Biography,* 12: 504–509.

13. Merricks, *World Made New,* p. 67.

14. Gillispie, *Dictionary of Scientific Biography,* 17: 143–48.

Chapter 20. The Quantum Revolution

1. Clifford E. Dykstra, *Physical Chemistry—A Modern Introduction* (Upper Saddle River, N.J.: Prentice-Hall, 1997), p. 333.

2. C. C. Gillispie, ed., *Dictionary of Scientific Biography,* 18 vols. (New York: Scribner, 1970–1990), 7: 379–82.

3. Ibid., 14: 337–42.

4. Ibid., 8: 551–52.

5. Ibid., 11: 7–17.

6. Ibid., 8: 180–83.

7. Lloyd Motz and Jefferson Hane Weaver, *The Story of Physics* (New York: Plenum, 1989), p. 245; Robert Resnick, "Misconceptions about Einstein," *Journal of Chemical Education* 57 (1980): 854.

8. Diana Kormos Barkan, *Walther Nernst and the Transition to Modern Physical Science* (Cambridge: Cambridge University Press, 1999), p. 115; Andrea Gabor, *Einstein's Wife: Work and Marriage in the Lives of Five Great Twentieth-Century Women* (New York: Viking, 1995).

9. Theoni Pappas, *Mathematical Scandals* (San Carlos, Calif.: Wide World Publishing, 1997), pp. 121–29.

10. Gillispie, *Dictionary of Scientific Biography*, 2: 239–54.

11. Ibid., 9: 606–607.

12. Ibid., 14: 597–99.

Chapter 21. Quantum Riddle

1. Max Tegmark and John Archibald Wheeler, "100 Years of Quantum Mysteries," *Scientific American* (February 2001): 68.

2. C. C. Gillispie, ed., *Dictionary of Scientific Biography*, 18 vols. (New York: Scribner, 1970–1990), 2: 487–89.

3. *The New Encyclopedia Britannica*, 12 vols. (Chicago: Encyclopedia Britannica, 1998), 2: 539–40.

4. Clifford E. Dykstra, *Physical Chemistry—A Modern Introduction* (Upper Saddle River, N.J.: Prentice-Hall, 1997), p. 332.

5. Gillispie, *Dictionary of Scientific Biography*, 3: 597.

6. Keith J. Laidler, *The World of Physical Chemistry* (New York: Oxford University Press, 1995), p. 334.

7. Walter Moore, *Schrödinger, Life and Thought* (New York: Cambridge University Press, 1989), pp. 63–66, 132, 295–96.

8. Gillispie, *Dictionary of Scientific Biography*, 12: 217–23.

9. D. Bergamini, ed., *Mathematics* (New York: Time, 1963), p. 69.

10. Gillispie, *Dictionary of Scientific Biography*, 12: 613–16.

11. Ibid., 17: 362–68.

12. Laidler, *The World of Physical Chemistry*, p. 331.

13. Ibid.

14. Gillispie, *Dictionary of Scientific Biography*, 15: 39–44.

15. Keith J. Laidler, *To Light Such a Candle* (Oxford: Oxford University Press, 1998), p. 311.

16. Ibid., pp. 321–23.

17. Paul Dirac, *Proceedings of the Royal Society* 123A (1929): 714.

Part V. The Flourish of the Physical Chemist

Chapter 22. Rebels and Radicals

1. *Grolier Encyclopedia of Knowledge,* J. Hacker, ed., 20 vols. (Danbury, Conn.: Grolier, 1993), 1: 326.

2. C. C. Gillispie, ed., *Dictionary of Scientific Biography,* 18 vols. (New York: Scribner, 1970–1990), 2: 132–34.

3. Ibid.

4. Keith J. Laidler, *The World of Physical Chemistry* (New York: Oxford University Press, 1995), p. 339.

5. Gillispie, *Dictionary of Scientific Biography,* 8: 289–94.

6. Richard N. Lewis, "A Pioneer Spirit from a Pioneer Family," *Journal of Chemical Education* 61 (1984): 3.

7. John W. Servos, "Lewis, the Disciplinary Setting," *Journal of Chemical Education* 61 (1984): 7.

8. Gerald E. K. Branch, "Gilbert Newton Lewis, His Influence on Physical-Organic Chemists at Berkeley, Appendix: Gilbert Newton Lewis, 1875–1946," *Journal of Chemical Education* 61 (1984): 21.

9. Gillispie, *Dictionary of Scientific Biography,* 8: 22–25.

10. Laidler, *World of Physical Chemistry,* p. 441.

11. Gillispie, *Dictionary of Scientific Biography,* 8: 473.

12. Laidler, *World of Physical Chemistry,* p. 420 (Heitler); Gillispie, *Dictionary of Scientific Biography,* 8 : 473–79 (London).

13. Donald A. McQuarrie, *Quantum Chemistry* (Mill Valley, Calif: University Science Books, 1983), p. 293.

14. Gillispie, *Dictionary of Scientific Biography,* 12: 418–20.

Chapter 23. Systems and Symmetry

1. C. C. Gillispie, ed., *Dictionary of Scientific Biography,* 18 vols. (New York: Scribner, 1970–1990), 10: 137–39.

2. Nina Byers, "E. Noether's Discovery of the Deep Connection between Symmetries and Conservation Laws," *Israel Mathematical Conference Proceedings,* vol. 12, 1999.

3. Victor J. Katz, *A History of Mathematics* (New York: Harper-Collins, 1993), p. 602.

4. Keith J. Laidler, *The World of Physical Chemistry* (New York: Oxford University Press, 1995), p. 351.

5. Tore Frängsmyr, ed., *Les Prix Nobel, 1998* (Stockholm: Almqvist & Wiksell International, 1998).

6. Gillispie, *Dictionary of Scientific Biography*, 14: 503–504.

7. Roald Hoffmann et al., *Chemistry Imagined: Reflections on Science* (New York: Smithsonian Institution Press, 1996).

8. Gillispie, *Dictionary of Scientific Biography*, 2: 169–71 (Bjerrum); D. Bergamini, ed., *Mathematics* (New York: Time, 1963), pp. 174–76 (Neumann).

9. Peter Atkins, *Physical Chemistry*, 6th ed. (New York: W. H. Freeman, 1998), p. 480.

10. Gillispie, *Dictionary of Scientific Biography*, 2: 83–85.

11. Thanks for the recollections go to Mary Tope, née Stone, Duke University, 1936.

12. Duke University Archives, Duke University. President William Preston Few, Records. Correspondence file.

13. Gerhard Herzberg, "Molecular Spectroscopy: A Personal History," *Annual Reviews of Physical Chemistry* 36 (1985): 1–30.

14. Donald A. McQuarrie, *Quantum Chemistry* (Mill Valley, Calif.: University Science Books, 1983), p. 379.

15. Herzberg, "Molecular Spectroscopy," 1–30.

16. Gillispie, *Dictionary of Scientific Biography*, 9: 433.

Chapter 24. The Ionists and Affinity

1. Diana Kormos Barkan, *Walther Nernst and the Transition to Modern Physical Science* (Cambridge: Cambridge University Press, 1999), p. 34.

2. C. C. Gillispie, ed., *Dictionary of Scientific Biography*, 18 vols. (New York: Scribner, 1970–1990), 1: 296–302.

3. Ibid., 6: 575.

4. George Wald, "How the Theory of Solutions Arose," *Journal of Chemical Education* 63 (1986): 658.

5. Gillispie, *Dictionary of Scientific Biography*, 10: 574–78.

6. Ibid., 10: 455–68.

7. Ibid., 2: 63–71.

8. Ibid., 4: 225–33.

9. Ibid., 6: 519–20.

Chapter 25. Intermolecular Forces

1. C. C. Gillispie, ed., *Dictionary of Scientific Biography*, 18 vols. (New York: Scribner, 1970–1990), 5: 273–403.

2. Ibid., 1: 160–61.

3. Ibid., 14: 109–11.

4. Maurice Rigby et al., *The Forces between Molecules* (Oxford: Clarendon Press, 1986), p. 1.

5. Gillispie, *Dictionary of Scientific Biography*, 3: 817–21.

6. Clifford E. Dykstra, *Physical Chemistry—A Modern Introduction* (Upper Saddle River, N.J.: Prentice-Hall, 1997), p. 106.

7. Geoffrey C. Maitland, *Intermolecular Forces—Their Origin and Determination* (Oxford: Clarendon Press, 1981).

8. Donald A. McQuarrie, *Quantum Chemistry* (Mill Valley, Calif.: University Science Books, 1983), p. 412.

Chapter 26. Heirs to the Ionists

1. John W. Servos, *Physical Chemistry from Ostwald to Pauling: The Making of a Science in America* (Princeton, N.J.: Princeton University Press, 1990), p. 66.

2. Ibid., p. 179.

3. Ibid., p. 94.

4. C. C. Gillispie, ed., *Dictionary of Scientific Biography*, 18 vols. (New York: Scribner, 1970–1990), 10: 328–29.

5. Richard J. Field and John A. Schellman, "Richard Macy Noyes," *Biographical Memoirs* 77 (1999): 224–45.

6. Paschalis Alexandridis et al., "A Record Nine Different Phases (Four Cubic, Two Hexagonal, and One Lamellar Lyotropic Liquid Crystalline and Two Micellar Solutions) in a Ternary Isothermal System of an Amphiphilic Block Copolymer and Selective Solvents (Water and Oil)," *Langmuir* 14 (1998): 2627.

7. Servos, *Physical Chemistry from Ostwald to Pauling*, p. 173.

8. Gillispie, *Dictionary of Scientific Biography*, 13: 521–24.

9. Ibid., 5: 492–95.

10. Karen Timberlake, *Chemistry*, 7th ed. (New York: Addison Wesley Longman, 1999), p. 609.

11. Servos, *Physical Chemistry from Ostwald to Pauling*, p. 303.

12. Timberlake, *Chemistry*, p. 273.

13. Servos, *Physical Chemistry from Ostwald to Pauling*, p. 314.

14. Diana Kormos Barkan, *Walther Nernst and the Transition to Modern Physical Science* (Cambridge: Cambridge University Press, 1999).

15. Ibid., p. 81.

Chapter 27. Reaction Rates

1. A. H. Zewail, "The Birth of Molecules," *Scientific American* (December 1990): 40–46.

2. C. C. Gillispie, ed., *Dictionary of Scientific Biography*, 18 vols. (New York: Scribner, 1970–1990), 14: 359–60.

3. Ibid., 6: 109–10 (Harcourt), 4: 411–12 (Esson).

4. Ibid., 9: 304–305.

5. Ibid., 3: 333–34.

6. Ibid., 2: 36–37.

7. Ibid., 18: 718–19 (Polanyi), 18: 279–83 (Eyring).

8. Ibid., 18: 279–83.

9. A. Maureen Rouhi, "The World of Isotope Effects," *Chemical and Engineering News* (December 22, 1997): 38–42.

10. Gillispie, *Dictionary of Scientific Biography*, 9: 620–24.

Part VI. The Fruits of Their Labors

Introduction

1. G. M. Eckert et al., eds., *Electropharmacology* (West Palm Beach, Fla.: CRC Press, 1990), p. i.

2. F. A. Miller and G. B. Kauffman, "C. B. Raman and the Discovery of the Raman Effect," *Journal of Chemical Education* 66 (1989): 795.

3. Clifford E. Dykstra, *Physical Chemistry—A Modern Introduction* (Upper Saddle River, N.J.: Prentice-Hall, 1997), p. 260.

4. Ulrich Burkert and Norman L. Allinger, *Molecular Mechanics,* ACS Monograph 177 (Washington, D.C.: American Chemical Society, 1982).

5. *Grolier Encyclopedia of Knowledge,* J. Hacker, ed., 20 vols. (Danbury, Conn.: Grolier, 1993), 6: 402.

6. *The New Encyclopedia Britannica*, 12 vols. (Chicago: Encyclopedia Britannica, 1998), 10: 142.

7. Ibid., 24: 72.

Chapter 28. Physical Chemistry and Physiology

1. Sinclair Lewis, *Arrowsmith* (New York: Harcourt Brace, 1925).

2. Ibid., p. 298.

3. Ibid., p. 405.

4. C. C. Gillispie, ed., *Dictionary of Scientific Biography*, 18 vols. (New York: Scribner, 1970–1990), 4: 614–16, 6: 410–12.

5. Ibid., 8: 445–46.

6. John W. Servos, *Physical Chemistry from Ostwald to Pauling: The Making of a Science in America* (Princeton, N.J.: Princeton University Press, 1990), p. 68.

7. Gillispie, *Dictionary of Scientific Biography*, 6: 423–25.

8. Arnold Thackray, ed., *Private Science: Biotechnology and the Rise of the Molecular Sciences* (Chemical Sciences in Society Series) (Philadelphia: University of Pennsylvania Press, 1998), p. 47.

9. Gillispie, *Dictionary of Scientific Biography*, 8: 50–53.

10. Ibid., 2: 397–402, 15: 61–64.

11. Joseph H. Noggle, *Physical Chemistry*, 2d ed. (Boston: Scott, Foresman and Company, 1989), p. 960.

12. Gillispie, *Dictionary of Scientific Biography*, 1: 319.

13. Ibid., 2: 16–30.

14. Ibid., 2: 17–19.

15. Georgina Ferry, *Dorothy Hodgkin: A Life* (Cold Spring Harbor, N.Y.: Cold Spring Harbor Laboratory Press, 1998).

16. Ibid., pp. 100–101.

17. Ibid., pp. 133–34, 241, 262.

18. J. M. Fenster, "The Conquest of Diabetes," *Invention & Technology* (winter 1999): 48–55.

19. Gillispie, *Dictionary of Scientific Biography*, 1: 440–43.

20. Ibid., 8: 614–15.

21. Ibid., 3: 351–53.

22. Fenster, "The Conquest of Diabetes," 48–55.

23. Louise S. Grinstein et al., eds. *Women in Chemistry and Physics: A Bibliographic Sourcebook* (Westport, Conn.: Greenwood Press, 1993), pp. 606–12.

24. George B. Kauffman and Laurie M. Kauffman, "An Interview with Linus Pauling," *Journal of Chemical Education* 73 (1996): 29–32.

25. Anne Sayre, *Rosalind Franklin and DNA* (New York: Norton, 1975), p. 151.

Chapter 29. Nonlinear Dynamics

1. Michael W. Young, "The Tick-Tock of the Biological Clock," *Scientific American* (March 2000): 61.

2. Anatol M. Zhabotinsky, "A History of Chemical Oscillations and Waves," *CHAOS* 1 (1991): 379–86.

3. James Prescott Joule, *Philosophical Magazine* 24 (1844): 106.

4. C. C. Gillispie, ed., *Dictionary of Scientific Biography*, 18 vols. (New York: Scribner, 1970–1990), 4: 556.

5. Ibid., 13: 253–54.

6. Keith J. Laidler, *The World of Physical Chemistry* (New York: Oxford University Press, 1995), p. 51.

7. A. M. Still, *The Soul of Amber* (New York: Murray Hill Books, 1944), p. 182.

8. M. W. Matlin, *Sensation and Perception* (Boston: Allyn and Bacon, 1988), p. 37.

9. Gillispie, *Dictionary of Scientific Biography*, 14: 199–201.

10. Ibid., 4: 199.

11. Sigmund Freud, *The Interpretation of Dreams,* trans. A. A. Brill (Las Vegas: World Library, 2000).

12. H. L. Heathcote, *Zeitschrift für Physikalische Chemie* 37 (1901): 368.

13. Zhabotinsky, "A History of Chemical Oscillations and Waves," pp. 379–86.

14. Bassam Z. Shakhashiri, *Chemical Demonstrations: A Handbook for Teachers of Chemistry (*Madison: University of Wisconsin Press, 1983).

15. Richard J. Field and John A. Schellman, "Richard Macy Noyes," *Biographical Memoirs* 77 (1999): 224–45.

16. E. S. Hedges and J. E. Myers, *The Problem of Physico-Chemical Periodicity* (New York: Longmans, Green, 1926).

17. Stephen K. Scott, *Oscillations, Waves, and Chaos in Chemical Kinetics* (New York: Oxford University Press, 1994), pp. 50–52.

18. James D. Murray, "How the Leopard Gets Its Spots," *Scientific American* (March 1988): 80–88.

19. Theoni Pappas, *Mathematical Scandals* (San Carlos, Calif.: Wide World Publishing, 1997), p. 63.

20. Jian Liu et al., *Nano Letters* 1 (2001): 57–60.

Chapter 30. Nanotechnology

1. Denis Brian, *Genius Talk: Conversations with Nobel Scientists and Other Luminaries* (New York: Plenum Press, 1995), p. 38.

2. Edward Regis, *Nano: The Emerging Science of Nanotechnology: Remaking the World—Molecule by Molecule* (Boston: Little, Brown, 1995).

3. Ibid., pp. 239–40, 250–51, 266, 272; Richard Smalley, "Of Chemistry, Love and Nanobots," *Scientific American* 285 (September 2001): 76.

4. Gary Stix, "Little Big Science," *Scientific American* 285 (September 2001): 32.

5. "Nanotechnology, A Special Report," *Chemical and Engineering News,* October 16, 2000, entire issue.

6. "Nanotech, Special Issue," *Scientific American* 285 (September 2001): entire issue.

7. Paul Alivisatos, "Less Is More in Medicine," *Scientific American* 285 (September 2001): 69.

8. Julia Karow, "Nanobending," *Scientific American* 285 (July 2000): 26.

9. Brian, *Genius Talk,* p. 11.

10. Charles M. Lieber, "The Incredible Shrinking Circuit," *Scientific American* 285 (September 2001): 59; Fenella Saunders, "Discover Dialogue: Mark Reed—Master of the Micro World," *Discover* 21 (September 2000): 26.

11. R. Dean Astumian, "Making Molecules into Motors," *Scientific American* 285 (July 2001): 57.

12. Tore Frängsmyr, ed., *Les Prix Nobel, 1998* (Stockholm: Almqvist & Wiksell International, 1998).

13. Philip Yam, "Profile: Richard E. Smalley; The All-Star of Bucky-Ball," *Scientific American* 269 (September 1993): 46.

14. Jessica Gorman, "A New Carbon Nanotool Springs to Life," *Science News* 58 (July 29, 2000): 71.

15. Phaedon Avouris, "Nanotubes for Electronics," *Scientific American* 283 (December 2000): 62–69.

16. "Richard Barry Bernstein," *National Academy of Sciences, Biographical Memoirs* 75 (1998): 38–57.

17. R. D. Levine, A. H. Zewail, and M. A. El-Sayed. "R. B. Bernstein Memorial Issue," *Journal of Physical Chemistry* 95 (1991): 7963.

18. Max Tegmark and John Archibald Wheeler, "100 Years of Quantum Mysteries," *Scientific American* 284 (February 2001): 72.

Chapter 31. Extreme Quantum

1. Max Tegmark and John Archibald Wheeler, "100 Years of Quantum Mysteries," *Scientific American* 284 (February 2001): 68–74.

2. Tore Frängsmyr, ed., *Les Prix Nobel, 1998* (Stockholm: Almqvist & Wiksell International, 1998).

3. Ibid.

4. Tegmark and Wheeler, "100 Years of Quantum Mysteries," pp. 68–74.

5. Ibid.

6. C. C. Gillispie, ed., *Dictionary of Scientific Biography*, 18 vols. (New York: Scribner, 1970–1990), 8: 473.

7. Clifford E. Dykstra, *Physical Chemistry—A Modern Introduction* (Upper Saddle River, N.J.: Prentice-Hall, 1997), pp. 490–91.

8. David Z. Albert, "Bohm's Alternative to Quantum Mechanics," *Scientific American* (May 1994): 58–67.

9. Tegmark and Wheeler, "100 Years of Quantum Mysteries," p. 68.

10. F. Selleri, ed., *Open Questions in Relativistic Physics* (Montreal: Apeiron, 1998), pp. 351–59.

11. F. David Peat, *Infinite Potential: The Life and Times of David Bohm* (Reading, Mass.: Addison-Wesley, 1996); Albert, "Bohm's Alternative."

12. Albert, "Bohm's Alternative."

13. Martin Gardner, "David Bohm and Jiddo Krishnamurti," *Skeptical Inquirer* 24 (2000): 20–23.

14. Tegmark and Wheeler, "100 Years of Quantum Mysteries."

15. Ibid.

16. Richard Wolfson, *Einstein's Relativity and the Quantum Revolution: Modern Physics for Non-Scientists,* part 1 (Springfield, Va.: The Teaching Co., 1999), tape 24.

17. Ibid.

Epilogue

1. C. C. Gillispie, ed., *Dictionary of Scientific Biography*, 18 vols. (New York: Scribner, 1970–1990), 18: 605–11.

2. A. Maureen Rouhi, "The World of Isotope Effects," *Chemical and Engineering News* (December 22, 1997): 38–42.

3. Ibid.

4. Robert K. Adain, *The Physics of Baseball*, 2d ed. (New York: Harper-Collins, 1994), pp. 12–15.

5. Gillispie, *Dictionary of Scientific Biography*, 18: 605–11.

6. Andrea Gabor, *Einstein's Wife: Work and Marriage in the Lives of Five Great Twentieth-Century Women* (New York: Viking, 1995), p. 150.

Selected Bibliography

HISTORIES OF PHYSICAL CHEMISTRY

Ihde, Aaron John. "Physical Chemistry I" and "Physical Chemistry II." In *The Development of Modern Chemistry*. New York: Harper & Row, 1964.

Laidler, Keith J. *The World of Physical Chemistry*. New York: Oxford University Press, 1995.

Nye, Mary Jo. *Before Big Science: The Pursuit of Modern Chemistry and Physics, 1800–1940*. Cambridge, Mass.: Harvard University Press, 1996.

———. *From Chemical Philosophy to Theoretical Chemistry: Dynamics of Matter and Dynamics of Disciplines, 1800–1950*. Berkeley: University of California Press, 1993.

Nye, Mary Jo, Joan L. Richards, and Roger H. Stuewer, eds. *The Invention of Physical Science: Intersections of Mathematics, Theology and Natural Philosophy since the Seventeenth Century. Essays in Honor of Erwin N. Hiebert*. Boston: Kluwer Academic, 1992.

Partington, James Riddick. "History of Physical Chemistry." In *A History of Chemistry*. 4 vols. New York: St. Martin's Press, 1961.

Servos, John W. *Physical Chemistry from Ostwald to Pauling: The Making of a Science in America*. Princeton, N.J.: Princeton University Press, 1990.

GENERAL REFERENCES

Atkins, Peter. *Physical Chemistry*, 5th ed. New York: Freeman, 1994.

Baptista Porta, John. *Natural Magick*. London: Printed for T. Y. Young and S. Speed, 1658.

Barkan, Diana Kormos. *Walther Nernst and the Transition to Modern Physical Science*. Cambridge: Cambridge University Press, 1999.

Bergamini, D., ed. *Mathematics*. New York: Time, 1963.

Bernal, J. D. *History of Classical Physics, from Antiquity to the Quantum*. New York: Barnes & Noble Books, 1972.

Brian, Denis. *Genius Talk: Conversations with Nobel Scientists and Other Luminaries*. New York: Plenum, 1995.

Brock, William Hodson. *The Norton History of Chemistry*. New York: W. W. Norton, 1993.

Bronowski, Jacob. *Magic, Science, and Civilization*. New York: Columbia University Press, 1978.

Brush, Stephen G. *The Kind of Motion We Call Heat: A History of Kinetic Theory of Gases in the Nineteenth Century*. Amsterdam: North Holland Publishing, 1976.

Cardano, Girolamo. *The Book on Games of Chance*. New York: Holt, Rinehart, and Winston, 1953.

———. *The Great Art*. Translated and edited by T. Richard Witmer. Cambridge, Mass.: MIT Press, 1968.

DuBos, Rene J. *Louis Pasteur: Free Lance of Science*. Boston: Little Brown, 1950.

Dykstra, Clifford E. *Physical Chemistry—A Modern Introduction*. Upper Saddle River, N.J.: Prentice-Hall, 1997.

Ferry, Georgina. *Dorothy Hodgkin: A Life*. Cold Spring Harbor, N.Y.: Cold Spring Harbor Laboratory Press, 1998.

Frängsmyr, Tore, ed. *Les Prix Nobel, 1998*. Stockholm: Almqvist & Wiskell International, 1998.

Gabor, Andrea. *Einstein's Wife: Work and Marriage in the Lives of Five Great Twentieth Century Women*. New York: Viking, 1995.

Gillispie, C. C., ed. *Dictionary of Scientific Biography*. 18 vols. New York: Scribner, 1970–1990.

Gouk, Penelope. *Music, Science, and Natural Magic in Seventeenth-Century England*. New Haven, Conn.: Yale University Press, 1999.

Grinstein, Louise S., et al., eds. *Women in Chemistry and Physics: A Bibliographic Sourcebook*. Westport, Conn.: Greenwood Press, 1993.

Hall, Bert S. *Weapons and Warfare in Renaissance Europe: Gunpowder, Technology, and Tactics*. Baltimore: Johns Hopkins University Press, 1997.

Hewitt, Paul G. *Conceptual Physics*, 11th ed. New York: Addison-Wesley, 1998.

Hewitt, Paul G., et al. *Conceptual Physical Science.* New York: Harper-Collins, 1994.

Katz, Victor. *A History of Mathematics.* New York: HarperCollins, 1992.

Laidler, Keith J. *To Light Such a Candle.* Oxford: Oxford University Press, 1998.

Larner, John. *Marco Polo and the Discovery of the New World.* New Haven, Conn.: Yale University Press, 1999.

Mann, John. *Murder, Magic and Medicine.* New York: Oxford University Press, 1992.

McQuarrie, Donald A. *Quantum Chemistry.* Mill Valley, Calif.: University Science Books, 1983.

Merricks, Linda. *The World Made New: Frederick Soddy, Science, Politics, and the Environment.* New York: Oxford University Press, 1996.

Moore, Walter. *Schrödinger: Life and Thought.* New York: Cambridge University Press, 1989.

Motz, Lloyd, and Jefferson Hane Weaver. *Story of Physics.* New York: Plenum, 1989.

Nye, Mary Jo. *Molecular Reality: A Perspective on the Scientific Work of Jean Perrin.* New York: American Elsevier, 1972.

Ostler, William. *The Evolution of Modern Medicine.* New Haven, Conn.: Yale University Press, 1921.

Pappas, Theoni. *Mathematical Scandals.* San Carlos, Calif.: Wide World Publishing, 1997.

Peat, F. David. *Infinite Potential: The Life and Times of David Bohm.* Reading, Mass.: Addison-Wesley, 1996.

Poirer, Jean-Pierre. *Lavoisier, Chemist, Biologist, Economist.* Translated by Rebecca Balinski. Philadelphia: University of Pennsylvania Press, 1996.

Rayner-Canham, Marelene, and Geoffrey Rayner-Canham. *Women in Chemistry: Their Changing Roles from Alchemical Times to the Mid-Twentieth Century.* New York: American Chemical Society, 1998.

Regis, Edward. *Nano: The Emerging Science of Nanotechnology: Remaking the World—Molecule by Molecule.* Boston: Little, Brown, 1995.

Rukeyser, Muriel. *Williard Gibbs.* Woodbridge, Conn.: Ox Bow Press, 1942.

Sayre, Anne. *Rosalind Franklin and DNA.* New York: Norton, 1975.

Scott, Stephen K. *Oscillations, Waves, and Chaos in Chemical Kinetics.* New York: Oxford University Press, 1994.

Skilling, Hugh Hildreth. *Exploring Electricity: Man's Unfinished Quest.* New York: Ronald Press Co., 1948.

Sobel, Dava. *Galileo's Daughter: A Historical Memoir of Science, Faith, and Love*. New York: Walker & Co., 1999.

Standen, Anthony. *Science Is a Sacred Cow*. New York: Dutton, 1950.

Stanley, Diane. *Leonardo da Vinci*. New York: Morrow, 1996.

Still, Alfred. *Soul of Amber: The Background of Electrical Science*. New York: Murray Hill Books, 1944.

Thackray, Arnold, ed. *Private Science: Biotechnology and the Rise of the Molecular Sciences* (Chemical Sciences in Society Series). Philadelphia: University of Pennsylvania Press, 1998.

Thorndike, Lynn. *A History of Magic and Experimental Science*. 8 vols. New York: Columbia University Press, c. 1923–1958.

Trevor-Roper, H. R. *The European Witch-Craze of the Sixteenth and Seventeenth Centuries, and Other Essays*. New York: Harper & Row, 1969.

Williams, L. Pearce. *Michael Faraday*. New York: Da Capo, 1987.

Wolfson, Richard. *Einstein's Relativity and the Quantum Revolution: Modern Physics for Non-Scientists*, part 1. Springfield, Va.: The Teaching Co., 1999.

Index

Abegg, Richard, 252
Aberdeen Proving Grounds, 370
Academy (Plato's), 37
Academy of Science, French, 72, 122, 124, 128, 209
acceleration, 118
acids, 49, 107, 309
affinity, 118, 219, 273, 275–83, 276
Africa, 29, 47, 50, 64
Africans, 46
afterimages, 339
Age of Exploration, 54
agriculture, 320
air
  bad, 126
  dephlogisticated, 126
  good, 126
  inflammable, 128
  very good, 126
alchemist, 21, 89, 111, 124
alchemy, 49, 59–60, 66, 109, 119
alcohol, 49
alcoholism, 302

Alexander the Great, 36, 43
Alexandria, Egypt, 47, 148
algebra, 26, 47, 64, 244, 296, 320
Algeria, 64
algorithm, 48
alpha helix, 334
alpha particle, 223
amalgam, 45
American Chemical Society, 302
American Medical Association, 302
American Physical Society, 355
American Revolution, 73
amino acid, 322, 334
Amistad, 181
ammonia, 184
ammonium chloride, 49
Amontons, Guillaume, 152
analgesic, 50
Anderson, John, 148
Andrews, Thomas, 288, 291
Anglican Church, 140
Anglo–Irish War, 105
Ångstrom scale, 323

Annals of Philosophy, 159
antacids, 104
antibacterial agents, 104
antidiarrheal, 50
antimony, 104
antiseptics, 50
Aquinas, Thomas, 56
Arab Empire, 21
Arabs, 26, 45–52, 54, 64, 112
Archimedes, 68
argon, 251–52
Argonne National Lab, 371
Aristarcus, 41
Aristotle 18, 20, 21, 22, 25, 35–44, 52, 58, 68, 73, 77, 78, 89, 94, 95, 105, 124–26, 139, 244, 374
Arnold, Winninette, 345
Arrhenius, Svante, 275, 278, 303, 311, 336
Arrhenius equation, 311
Arrowsmith, Martin, 325
Arrowsmith (Lewis), 325
Astbury, William, 329
astrolabe, 49
astronomy, 28
astrophysics, 320
atom, 18–19, 149, 190, 217
atomic orbital, 267
atomic spectra, 237
atomic structure, 217–27
atomic weights, 297
atoms, 69, 139, 239, 248, 374
Auschwitz, 310
Avogadro, Amedo, 187
Avogadro's number, 186–87, 192, 194

Babylonians, 30

Bacon, Francis, 198
Baghdad, 48
baking soda, 107, 129, 297, 307
balance, 49, 124
balloonists, 290
Balmer, Johann, 206, 227, 238, 243
Bancroft, Wilder, 295, 298, 300, 333
Banting, Frederic, 331–32
Barrow, Isaac, 111, 113, 116
bases, 107
base-sixty system, 47
batteries, 303
Becquerel, Henri, 221
belladonna, 56
Bell, John, 365
Belousov, Boris, 344–45
Belousov–Zhabotinsky (BZ) reaction, 344, 346, 347
bends, 141
Benjamin of Tudela, 51
Bernal, John D., 54, 329–31, 333, 352
Bernard, Claude, 302
Bernoulli
  Daniel, 157, 160, 165
  family, 157–58
  Johann, 158
Bernstein, Richard, 355
Bernstein's wiggles, 355
Berthelot, Marcellin, 20, 283, 304
Berzelium, Jöns Jakob, 144
beta particle, 223
bibliographic ghost, 338
Bigeleisen, Jacob, 315–16
Bigollo, 64. *See also* Fibonacci
Binnig, Gerd, 324
biological clocks, 337
biology, 32

Birge, Raymond, 271
Bismarck, Otto von, 196
Bjerrum, Niels, 270
black art, 87
blackbody, 229–30, 233
blackbody radiation, 289, 360
Black Death, 54
blood, 105
blood glucose, 331
Bodenstein, Max, 311–13, 314
Bohm, David, 364–65
Bohr, Niels, 227, 228, 237, 239, 242, 250, 252, 357, 363
Boisbaudran, Lecoq, 206
Boltzmann, Ludwig, 30, 155, 157, 165, 166, 172–76, 184, 186, 228, 230, 244, 270, 279, 289, 305, 315, 360
Boltzmann's constant, 175
bomb calorimeter, 304
bond, 255
bond length, 321
Book on Games of Chance (Cardano), 167
Borgia, Cesare, 66
Born, Max, 20, 244, 245, 369
Born–Oppenheimer approximation, 256
Bose, Satyendranath, 360, 361
Bose–Einstein condensate, 360
bourgeoisie, 122, 128
Boyle, Robert, 78, 101, 105–109, 119, 124, 131, 139, 157, 158, 290, 337, 374
Boyle's law, 108, 288
Bradley, James, 41
Bragg, William Henry, 328, 333
Bragg, William Lawrence, 259, 328, 333

Brahe, Tycho, 114
Bray, William, 342–43
Brescia, Italy, 81
Brewster, David, 201–202
bricks, 26
Broglie, Louis de. *See* de Broglie, Louis
Brown, Robert, 188–89
Brownian motion 188, 193
Bruno, Giordano, 96
Bryn Mawr College, 263
bubonic plague, 54
buckminsterfullerine, 352
Bucky balls, 352, 355
Bunsen, Robert, 202–204, 229, 230
Burbury, S. H., 369
Bury, Charles, 252
Byzantine Empire, 46
Byzantium, 64

calcination, 125
calcium carbonate, 104, 107, 219
calcium chloride, 219
calculus, 30, 78, 297, 320
California Institute of Technology. *See* Cal Tech
Callisthenes, 43
caloric, 150, 287
calorimeter, bomb, 304
Cal Tech, 258, 259
Cambridge Univeristy, 136, 140, 329
Camera Obscura, 196
cannonball, 71, 91
carbon, 264, 353
carbonated soda, 337
Cardano, Giolamo, 81, 82–86, 87, 93, 166–67, 243, 374
Carnot, Sadi, 153

catalysis, 311
cathode, 220
cathode ray, 220, 362
cathode ray tube, 217, 220–21, 241
Catholic, 89
Catholic Church, 56, 75
Celeste, Maria, 99–102
Celtic tales, 54
censo, 87
cesium, 204
Chadwick, James, 227
chain reaction, 313
chalk, 219
Chandoux, 89
Châtelier, Henry le, 184
chemical bond, 250, 256, 257
chemical bonding, 247
chemical equilibrium, 183
chemical philosophy, 247
chemistry, 74, 78, 112, 177
Chemistry, Benevolent and Easy, Favorable to Women, The (Meurdrac), 103
childbirth, 56, 59, 104
China, 47, 71
Chinese, 46, 57
chloride ion, 251
Christians, 52, 46
Christian religion, 43, 54
Church of England, 136, 144
classical mechanics, 239
Clausius, Rudolf, 153–55, 164, 169, 179, 278, 374
coal, 52, 133
Cohen, Ernst, 310
Cohn, Edwin, 327, 332
cold packs, 179
Collection of Short Papers on Radiation Medicine, A (Belousov), 344

Collip, James, 332
Collip's extract, 332
colloid, 299, 333
Columbia University, 370
combustion, 127, 131, 304
communist, 334
complementarity, 242
compressible atoms, 297
computer, 320, 322
Comte, Auguste, 319
conservation of mass, 125, 126
Conversations on Chemistry (Marcet), 211
Copernican system, 96
Copernicus, Nicolaus, 96, 112, 114, 144
copper, 224
Cornell, Eric, 361
Cornell University, 298
corpuscle, 139
cosa, 86, 87
cosmetics, 77
Coulomb, Charles, 251
Coulomb attraction, 251
Coulomb interaction, 285
Coulomb's law, 251
Crick, Francis, 334–35
Crookes, William, 223
Crusades, 50
cubic equation, 82
cubo, 87
Curie
  Marie, 221
  Marie and Pierre, 190
Curl, Robert, 353
cyclol bond, 333

Daguerre, Louis, 196
Dalton, John, 22, 23, 132, 136,

138–46, 149, 151, 157, 161, 185, 230, 287, 374

Darwin, Charles, 134

da Vinci, Leonardo, 22, 23, 54, 57, 65–71, 287

Davisson, Clinton, 241–42

Davy, Humphry, 159, 211, 218–19, 250

de Broglie, Louis, 239–41, 244, 364

Debye, Peter, 291, 294, 309, 321

decimal fraction, 88

decoherence, 365

deduction, 30–31

della Porta, Giambattista, 53, 57–61, 67, 76, 78, 103, 104, 196, 374

de Méré, Chevalier, 167

Democritus, 139

denaturing, 332

density-functional theory, 359–58

deoxyribonucleic acid. *See* DNA

Descartes, René, 88–91, 152

diabetes, 331

dialectic, 38

Dialogue on the Two Chief Systems of the World (Galileo), 98

diamonds, combustion of, 127

diethyl ether, 264

differentiation, 113

diffraction, 208–209

dilithium crystals, 272

Dingle, Herbert, 205

dipole, temporary, 286

dipole interaction, 286, 291, 321

Dirac, Paul, 345, 359, 363

directed valence, 260

Discourse on Method (Descartes), 91

Discourses and Mathematical Demonstrations Concerning Two New Sciences (Galileo), 98

dispersion forces, 293

DNA, 334–35, 351

Don Quixote (Cervantes), 48

Drexler, Eric, 329, 354

drug addiction, 302

Du Châtelet
    Marine de Bertereau, 102
    Marquise, 121

Duhem, Pierre, 283

Duke University, 258, 271

Dulong, Pierre, 143–44, 235

Du Pont, 124

E. I. du Pont de Nemours & Company, 124
    Eleuthère, 124
    Pierre Samuel, 124, 131

Dürer, Albrecht, 69, 70

dysprosium, 206

Easter, 64

East India Company, 162

eclipses, 40

Egypt, 43, 64

Egyptians, 46

Ehrlich, Paul, 316

Einstein, Albert, 20, 30, 205, 228, 233–34, 235–37, 241, 244, 245, 263, 301, 306, 313, 320, 359, 360, 361, 363

electric field, 243

electricity, 277

electrochemical cell, 303

electrochemistry, 218, 276, 277

electrode, 220

electrodynamics, 229, 247

electromagnetic field, 269

electromagnetic wave, 374

electromagnetism, 207–15, 360

electron orbits, 256

electrons, 217–18, 220, 239, 241, 252, 255
electron spin, 255, 372
electrostatic energies, 257
electrostatic interaction, 251
Elementary Principles in Statistical Mechanics (Gibbs), 176, 185
Elementary Treatise on Chemistry An (Lavoisier), 128
elements, 32, 42
Elements of Chemistry (Lavoisier), 140
Elements of Psychophysics (Fechner), 342
energy, 66, 149, 178, 179, 304, 373
English Civil War, 105
Enlightenment, 125
entanglement, 362, 365
Enterprise, 272
entropy, 153–55, 172, 174, 179, 183, 280, 304, 344, 373, 374
enzymes, 287, 336
epicycles, 33
equilibrium, 177, 179, 180, 184, 247
equilibrium constant, 276, 315
equipartition theorem, 162–63
equivalents, 144, 190, 219
Eratosthenes, 41
ergot, 56
Esson, William, 309
ethanol, 264
ether, 212
Euclid, 32, 42–43, 80, 82, 86, 112
Eudoxus of Cnidus, 41
Everett, Hugh, 365
exchange energy, 257
Extraordinarius, 136
Eyring, Henry, 24, 313–15

Faraday, Michael, 135, 207, 211–14, 218–19, 223, 250, 339, 374
Fechner, Gustav, 134, 337–42
Fermi, Enrico, 372
Ferro, Scipione, 83
Ferry, Georgiana, 330
fetus, 108
fever, 50
Few, W. P., 271
Feynman, Richard, 239, 348–49, 360, 367
Fibonacci, Leonardo, 64, 67, 86
field, 212–13, 374
Fiore, Antonio Maria, 83
folk remedies, 50
force, 28, 66, 68, 118, 219, 250, 289
force at a distance, 118
Foresight Institute, 349
four elements, 124, 126
four-element theory, 39
Fourier, Jean, 321
Fourier transform spectroscopy, 321
Franck, James, 271
Franklin, Rosalind, 334–35
Fraunhofer, Joseph von, 200–201
Fraunhofer lines, 201
free energy, 304 344, 374
freezing point, 325
French Revolution, 73, 128–31, 240, 287
Fresnel, Augustin, 209–10
Freud, Sigmund, 342
friction, 68, 373
fugacity, 293
Fukui, Kenichi, 267
Fulhame, Elizabeth, 196
Fuller, Buckminster, 353

function, 91
wave, 257

Galen, 42–43, 112
Galilei
Galileo. *See* Galileo
Vicenzo, 93
Galileo, 20, 30, 36, 40, 78, 89, 91–100, 102, 108, 112, 118, 231, 239, 244
Galileo's Daughter (Sobel), 100
gallium, 206
Galois, Evariste, 30, 262, 264, 305
Galvani, Luigi, 210
Gamba, Marina, 99
gambling, 84
gamboge, 191, 196, 301
gas effusion, 161
gas law, 280, 374
Gay-Lussac, Joseph, 158
Geiger, Hans, 217
general relativity, 263
gentleman scholar, 125, 207, 296
gentlewoman scholar, 125
geocentric theory, 39
geology, 320
geometry, 28, 86, 91
Germain, Sopie, 264
German–French War of 1870, 155
Gibbs, Josiah Williard, 176, 177–85, 186, 276, 283, 305, 319, 342, 374
Gibbs free energy, 183
Gibbs's phase rule, 298
glory effect, 355
glow-in-the-dark materials, 221
glycolysis, 345
gold, 49–50
Goldstein, Eugene, 220, 230

Göttingen University, 263, 291
Gottleib, Max, 325
Goudsmit, Samuel, 243, 363
Graham, Thomas, 159, 301
Grand Unified Theory (GUT), 214, 367
Grant College, 162
graphite, 352
gravity, 28, 40, 91, 112, 114, 118, 219–20, 275, 284, 287, 374
gray goo problem 354
Greek classics, 57
Greeks, 26, 27–34, 46, 64, 88, 93, 112, 210
ancient, 21
Droian, 29, 30
group theory, 262, 264, 270, 329, 370
guillotine, 130, 321
Guldberg, Cato, 178, 276
gunpowder, 58, 71, 77, 104, 123

Haber, Fritz, 184
Halley, Edmond, 116
hallucinogenics, 56
Harcourt, Vernon, 309
Harmonies of the World (Kepler), 116
Harvard Medical School, 327
Harvard University, 254, 255, 268, 296, 298, 345
Haverford College, 296
Hawking, Stephen, 117
heart, 337
heat, 147–56, 308, 344
heat capacity, 163, 235
heat capacity ratio, 175, 176
heating, 151
conductive, 151

radiative, 151

Heisenberg, Werner, 243–44, 245, 257

Heisenberg's uncertainty principle, 257

Heitler, Walter, 250, 256–57, 272

Helios, 205

helium, 161, 205, 251–52, 254, 293

Helmholtz, Hermann von, 152–53, 164, 283, 324, 327, 342

hemoglobin, 332

Henry, William, 141

Henry's Law, 141

Herapath, John, 158–59, 163, 164, 165, 288

herbs, medicinal, 42, 104

Herschback, Dudley, 354

Herschel
  Caroline, 168, 199
  John, 168, 199, 201
  William, 168, 199

Hertz, Heinrich, 215, 234, 238

Herzberg, Gerhard, 272–73

Hindus, 46, 48

Hipparchus, 32

Hippasus, 88

History of Animals (Aristotle), 37

Höber, Anselm, 327

Hodgkin, Dorothy, 330–31

Hoff, Jacobus, van't. *See* van't Hoff, Jacobus

Hoffmann, Roald, 268–69

Holy Roman Empire, 46

Homo sapiens, 134

Horstmann, August, 284

House Committee on Un-American Activities, 335, 364, 366

Hückel, Eric, 294

hydrocarbons, 251

hydrogen, 138, 184, 251–52, 256, 264, 286, 288, 289

hydrogen bonding, 332–33

hydrogen chloride, 313

hydrogen iodide, 313

hydrogen peroxide, 342

Hypatia, 44

ibn-Munqidn, Usama, 50

ideal gases, 288

Iijima, Sumio, 353

imaginary number, 243

India, 47

indicator, 107

induction, 30–31

inductive proof, 99

Industrial Revolution, 22, 52, 133–35, 151, 197

inertia, 99

infrared light 200, 321

inheritance powders, 56

inorganic compounds, 131

Inquisition, 54, 56, 59, 85, 89, 98

insanity, 302

insulin, 331

integration, 113

interference, 208, 355, 360

intermolecular attractions, 118

intermolecular forces, 249, 285–94, 353, 373

intermolecular interactions, 309

Interpretation of Dreams, The (Freud), 342

Invisible Man, The (Wells), 201–202

ioner, 273

ionists, 248–49, 273, 275–83, 294–95, 307, 308, 309, 327, 336, 345

ion pump, 352
ions, 248, 273, 278, 281
iron, 342
Islamic Civilization, 45
Islamic empire, 46
isotopes, 225, 371

Al-jabr wa'l muqábalah, 48
Japanese Americans, 260
Jews, 46, 56
Johns Hopkins University, 369–70
Jordan, Michael, 262
joule (unit of energy), 151
Joule, James, 135, 151–52, 153, 164, 289, 337
Joule-Thomson effect, 289
Journal of Chemical Physics, 314
Journal of Physical Chemistry, 298, 301
journals, 135
Julius, William, 205–206
Jupiter, 95

Kaiser Wilhelm Institute for Physical Chemistry, 314
kaleidoscope, 201
Al-Khazini, 49
Al-Khwarizmi, 48
Kelvin, Lord, 207, 238, 280
Kepler, Johannes, 112, 116
Khan, Genghis, 51
kinetic molecular theory, 186, 190
kinetic theory of gases, 136, 157–65, 168–72, 190, 193, 270, 286, 288, 301, 307, 313
King's College, 335
Kirchhoff, Gustav, 202–204, 229–31
Koch, Robert, 134

Kohn, Walter, 357–59
Kowa, Seki, 116
Krishnamurti, Jiddo, 365
Krönig, Karl, 164
Kroto, Harry, 353
krypton, 251
Kundt, August, 175
Kyayyám, Omar, 48

Langevin, Paul, 241
Langmuir, Irving, 255–56, 333
Laplace, Pierre-Simon, 168–69, 287, 288
laser, 208, 320, 329, 361
latitude, 76
Laue, Max, 328
Lavoisier
   Antoine, 65, 78, 103, 120–32, 133, 135, 138, 139–40, 141, 143, 150, 168, 189, 196, 211, 239, 374
   Marie, 123, 125, 128–30, 150
law of mass action, 178, 180, 278, 297, 308, 336
law of multiple proportions, 140
Lee, Yuan T., 354
Leibig, Justus, 144
Leibniz, Gottfried, 116–17
Leipzig City Advertiser, 195
Leipzig University, 340
lemon battery, 278
Lenard, Philipp, 234
Lennard-Jones, John, 266, 272
lenses, 60
Leonardo of Pisa, 64–65
Leopard's spots, 346
Leucippus, 139
lever, 68
Lewis, G. N., 10, 250, 252–56, 293,

302, 303, 314, 319, 333, 342, 369, 374
Lewis, Sinclair, 325, 326
Lewis dot structures, 261
light, 269
　particle model, 235
　scattering, 321
　speed of, 214
　wave model, 208–209, 235
limonene, 144
linear equation, 82
Linné, Carl von. *See* von Linné, Carl
Lippmann, Gabriel, 196
liquid-drop model, 372
lithium, 205
Lockyer, Joseph, 205, 251
lodestone, 54, 59
Loeb, Jacques, 326–27, 339
logarithms, 88
London forces, 293
London, Fritz, 256–58, 291
London Times, 159
longitude, 76
Lonsdale, Kathleen, 329
Lorentz, Hendrik, 228
Loschmidt, Josef, 186–87
Lucasian professor, 117
Lucretius, 285
Lummer, Otto, 230
Luther, Martin, 87
Lyceum, 37

Mach, Ernst, 144, 194
machines, war, 66
Macleod, John, 332
magician, 21, 26, 27, 53, 96
magick, 74
magic, natural, 57–61
magic numbers 372

magnet, 353
magnetic field, 243, 272, 255
magnetism, 59, 210
Manhattan Project, 260, 372
Marcet, Mrs., 211
Marić, Mileva, 234
marrow, 105
Massachusetts Institute of Technology. *See* MIT
mass action, 276
mass spectrum, 353
mathematical chemistry, 301
mathematicians, 77
mathematicians, Renaissance, 112
mathematics, 28, 64, 78
　Babylonian, 21
　Byzantium, 21
　Egyptian, 21, 30
Maxwell
　James Clerk, 165, 169–76, 179, 183, 184, 189, 190, 207, 214, 234, 235, 244, 270, 289, 360, 367, 374
　Katherine, 169
Maxwell's demon, 165, 172, 352
Mayer
　Joseph, 369
　Marie, 315–16, 369–74
mean-free path, 165
mechanical engineer, 53
mechanical universe, 133
mechanics, 39, 66
mechanisms, 310, 373
medicine, 320
　Arabic, 50
medicines, 58
Mendeleev, Dmitri, 20, 134, 202, 224, 251–52
Menschutkin, Nikolai, 309

Menten, Maud, 316
merchants, 80
mercury, 45, 104, 119, 126, 175
Méré, Chevalier, de. *See* de Méré,
    Chevalier
Merlin, 54
metabolism, 131
meteorology, 39
methane, 260
metric system, 103
Meurdrac, Marie, 102, 127
Meyer, Lothar, 204
Michaelis, Leonor, 316
Michaelis–Menten model, 316
microscope, 192, 201, 223, 243,
    323
microwave radiation, 321
Middle ages, European, 53, 58, 63
midwives, 56, 59
Millikan, Robert, 271
misease, 339
Mises, Dr., 339
MIT, 254, 268, 297
mole, 187
molecular beams, 354
molecular modeling, 322
molecular orbital, 267
molecular orbital theory, 266–67,
    268
molecules, 69, 99, 149, 187
momentum, 241
Mona Lisa, 66
Mongols, 51
Moscow State University, 345
Moseley, Henry, 224, 251
motion, 39, 92, 94
movable type, 57
Mulliken, R. S., 266
Munqidn. *See* ibn-Munqidn

Musa, Mansa, 50
Muslims, 48, 56
myoglobin, 301

Nagaoka, Hantaro, 237
nanobalance, 351
nanobearings, 353
nanogears, 351
nanometer, 349
nanomotors, 351, 352
nanoparticles, 350
nanoscale, 347, 350
nanotechnology, 320, 348–55, 373
nanotransistors, 351
nanotubes, 353
nanovessels, 351
nanowater, 350
nanowrenches, 351
Napier, John, 88
Napoleon, 209
National Socialists, 291
natural magic. *See* magic, natural
Natural Magick (della Porta), 57–
    61
natural philosophy, 74
nature, 205, 290
Nazis, 245, 258, 263, 291, 304
negative numbers, 65
neon, 251–52, 293
Nernst
    Emma, 304, 327, 336
    Walther, 303–304
nerve cells, 302
Neumann, John von, 270
neutrons, 217, 218, 227, 372
Newcomen, Thomas, 148
*New System of Chemical Philosophy,
    A* (Dalton), 139, 146
Newton, Isaac, 18, 30, 40, 78, 109–

20, 132, 133, 152, 157, 158, 162, 166, 168, 198, 199, 209, 212, 223, 231, 239, 244, 284, 287, 288, 374

Newton effect, 134

nickel, 224, 241

Niepce, J., 196

nitrates, 184

nitric acid, 342

nitrogen, 161, 184, 251, 322
  gas, 141
  liquid, 286

Nobel Prize, 217, 220, 221, 223, 224–25, 227, 234, 237, 238, 242, 269, 271, 291, 297, 303, 316, 321, 328, 331–32, 334, 344, 353, 354, 355, 357, 358, 373

Noether, Emmy, 30, 262–64, 271, 291, 305, 362, 374

nonlinear dynamics, 320, 336–47

Noyes
  Arthur Amos, 297, 345
  family, 298
  Richard, 345, 346
  William, 345

nuclear shell model, 372

Nuclear Test Ban Treaty, 334

nucleus, 241

numerals
  Hindu, 46
  Indian, 64
  Roman, 46, 64

Oersted, Hans, 210

Oettinger, Luise, 273

"On Faraday's Line of Force" (Maxwell), 214

optics, 201

orbit, 259

orbital angular momentum, 372

orbitals, 266, 306–307

orbits
  atomic, 262
  molecular, 262

Order of the British Empire, 346

Oregonator, 346

Oregon State University, 258

organic chemistry, 269, 296

organic compounds, 131

Origin of the Species (Darwin), 134

oscillations, 337–47, 374

osmosis, 276, 279–80

osmotic pressure, 279–80, 324, 328

Ostwald, Wilhelm, 17, 187, 193, 278–79, 281, 297, 303, 309, 311, 319

Oxford Univeristy, 136, 140, 330, 331, 333

oxygen, 127, 138, 251, 264, 272, 286, 322

Pacioli, Luca, 65

pagan, 54

pancreas, 331

paper, 47

Paracelsus, 103, 127

paradigm, 20, 174

parchment, 43

Parsons, Cooke, 296

Pascal, Blaise, 167

Passover, 64

Pasteur
  Louis, 134
  Marie, 125

Pauling, Linus, 258, 272, 317, 319, 333, 364

penicillin, 331

Pennsylvania State College, 364
periodic law, 20
periodic table, 204, 224, 252
peripatetic, 37
permanent gas, 291
perpetual motion, 67
Perrin, Jean, 185–94, 196, 227, 228, 237, 301, 311, 322, 374
perspective, 69
Peterson, Chris, 349
Petit, Alexis, 143, 235
Pfeffer, William, 280
phase rule, 298
Phase Ruler, 299
Philip II, 36
Philippines, 254
philosophy, 29
phlogiston, 125–26
phosphorescing, 221
photochemistry, 254, 313
photoelectric effect, 234
photography, 196, 221
photons, 313
physical chemistry, 13, 17, 19, 39, 189, 247, 276, 296–97
Physical Chemistry of the Cell and Tissues (Nernst), 327
physical chemists, 248
physics, 74, 78
physiology, 316, 320, 325–28
pilot waves, 364
pirates, 76
plague, 54, 55, 58, 75, 98, 102, 105
Planck, Max, 228, 230–34, 235, 250, 270, 289, 303, 305, 320
Planck's constant, 314
Plato, 25, 27, 32, 35, 205
plum pudding model, 221, 224, 252

Poisson, Siméon, 209
Polanyi
   John, 354
   Michael (Mihály), 313–14
Polo, Marco, 51–52
polytheists, 46
Pople, John, 357–59
Porta, Giambattista della. *See* della Porta, Giambattista
positional notation, 47
potassium, 211, 219
potential energy surfaces, 314
pressure, 99
Prester, John, 51
Priestley, Joseph, 126, 339
Prigogine, Ilya, 343, 346
Princeton Institute for Advanced Study, 263
Princeton University, 364, 365
Principia (Newton), 116, 121, 122
prism, 199, 322
privatdozent, 136, 263
probabilities, 167–68
probability, 174, 256
probability waves, 244
Prohibition, 370
proteins, 301, 332, 334
Protestant Reformation, 22, 73, 196
Protestants, 89
protons, 217–18, 372
psychology of sensation, 339
psychophysics, 342
Ptolemy, 32, 112
purgatives, 101
purge, 50
Puritans, 297
Pythagoras, 30, 88, 93, 112
Pythagoreans, 88
Pythagorean theorem, 30

quadratic equation, 82
quantum, 228, 241
  chemistry, 250, 286, 301
  chemists, 248–49
  dots, 351
  electrodynamics, 355, 367
  mechanics, 164, 196, 229, 239–45, 247, 250, 313, 315, 320, 357–67, 370
    Copenhagen interpretation, 256
    origin of term, 244
  model, 237
  physics, 250
  weirdness, 359, 374
quinine, 50

radiation hypothesis, 311
radiative heating, 229
radicals, 254
radioactivity, 221
radiochemistry, 370
radiometer, 223
radio receiver, 215
radon, 251
Railway Magazine, 159
rainbow scattering, 355
Raman, Chandrasekhara, 321
Ramsay, William, 152, 205
ratcheting, 352
rate constant, 315
rate of reactions, 83, 306–17, 373
Rathenau, Emil, 303
Rayleigh, Lord, 163, 165
reactivity, 261
reducio ad absurdum, 30–31, 94
Reformation, Protestant, 54, 75, 140
Reimann, 29

remedies, 101
Renaissance, 22, 26, 53, 63, 77
Research Laboratory of Physical Chemistry, 298
rest mass, 218
Restoration, 209
reversibility paradox, 174
Revolutions of the Heavenly Bodies (Copernicus), 96
rhetorical, 86
Rice, Francis, 354
Richards, Theodore W., 254, 255, 295–97, 298
Richardson, Lewis, 285
Richter, Jeremias, 129, 140
Ritter, Johann, 200
Roberts, Margaret, 331
Robespierre, 321
Robinson Crusoe (Stevenson), 254
Rockefeller Institute for Medical Research, 317
Rohrer, Heinrich, 324
Roman Catholic Church, 22. *See also* Catholic Church
Romans, 21, 26, 46
Rome, 43
Röntgen, Wilhelm, 221, 328
rotational motion, 148–50
Royal Society, 107–108, 116, 119, 159, 163, 169, 188, 329
rubidium, 204
Rumford, Count. *See* Thompson, Benjamin
Ruska, Ernst, 323
Rutherford, Ernest, 221–24, 225, 227, 228, 237

Sala, Angelo, 196
sal ammoniac, 49

samarium, 206
Santayana, George, 25
scanning electron microscope, 323
Scholasticism, 52, 56–57, 71
Scholastics, 54, 56
Schrödinger, Erwin, 242–43, 244, 245, 260, 264, 359, 361, 374
Schrödinger equation, 266
Schrödinger's cat, 361, 362
Schweigger, Johann, 338
Schweigger's Journal for Chemistry and Physics (Fechner), 338
Schwinger, Julian, 367
scientific method, 38
Scientific Revolution, 22, 26, 56, 71, 73–132
sea trade, 76
sea urchin, 327
semiconductors, 357
serum albumin, 327, 332
Sforza, Ludovico, 66
shampoo, 299
Shelburne, Lord, 126
Sheldon, Mary, 254
sickle cell disease, 317
siderism, 200
Sidgwick, Nevil, 260
silver cyanate, 144
silver fulminate, 144
single-molecule sensor, 351
skin lesions, 50
skull, 105
Slater, John, 260
Smalley, Richard, 353
smuggling, 123
Sobel, Dava, 100
Soddy, Frederick, 225–27
sodium, 204–205, 211
sodium bicarbonate, 129, 297

sodium chloride, 251
sodium ion, 251
sodium oxide, 218
sodium thiocyanate, 302
Sorbonne, 190
sorcerers, 27
sorcery, 57–58
sound, 235
spark plug, 215
spatial oscillations, 346
spectroscope, 204
spectroscopists, 198
spectroscopy, 198–206, 215, 229, 261, 269, 320, 357, 370
spectrum, 199
spin coupling, 372
spins, 255
Sponer, Hertha, 271
Standen, Anthony, 18
Stark, Johannes, 243, 245
Starry Messenger, The (Galileo), 95
static electricity, 219
statistical mechanics, 166–76, 174, 261, 313, 315, 370
statistics, 256
Stevin, Simon, 88
Stifel, Michael, 87
stoichiometry, 139
Stoney, George, 220
structural engineering, 320
Struven, Mildred Witte, 147
sucrose, 309
superconductivity, 361
superposition, 365
Swamp gas, 260
symmetry, 30, 262–63, 264, 266, 268, 270, 329, 370, 374
syphilis, 316
Syria, 44, 64

Tait, Peter, 172, 183
Talbot, William 201
Tartaglia, Nicolò, 81–86, 87, 93
tax farm, French, 123, 129
telescope, 61, 71, 76, 95, 112, 201
temporary dipole. *See* dipole, temporary
Thales, 29
Thatcher, Margaret, 331
theoretical chemistry, 247
thermodynamics, 20, 52, 147–56, 177–85, 229, 297, 313, 315, 344, 370
  first law, 150–56
  second law, 150–56
  third law, 304
Thirty-nine Articles of Religion, 136
Thirty Years War, 22, 73, 98, 196
Thomson
  Benjamin (Count Rumford), 131, 150
  George Paget, 242
  Joseph John, 217, 220–22, 223, 227, 234, 237, 242, 252, 289, 328, 353, 362
  William (Lord Kelvin), 133, 151–52, 153
thorium, 223
Thoughts on Mental Functions (Waterston), 163
three-body problem, 256
time, 302
tin, 126
tobacco, 129
Tomonaga, Shinichiro, 367
Transactions of the Connecticut Academy, 183
transition state, 314
translational motion, 148–50

transmission coefficient, 315
transmission electron microscope, 323
transmutation, 125, 223
trebucher, 75
trigonometry, 28, 49
Trinity College, 113
tropism, 326
Tsai-Yü, Chi, 93
tunneling, 323–24, 359
Turing, Alan, 346
Turks, 21
two-electron bond, 254, 374
Tyndall, John, 301

ultraviolet catastrophe, 230
ultraviolet light, 200
ultraviolet radiation, 321
uncertainty principle, 243–44, 245
University of Adelaide, 328
University of Amsterdam, 279
University of Berlin, 303, 316, 326
University of Calcutta, 360
University of California at Berkeley, 271, 364
University of California at San Diego, 373
University of Chicago, 371, 373
University of Heidelberg, 365
University of Leiden, 290
University of Nagoya, 316
University of Saskatchewan, 273
University of Uppsala, 278
University of Zürich, 155
uranium, 221, 223, 315
Uranus, 199
Urban VIII, Pope, 96
Urey, Harold, 354, 370
uterus, 337

vacuums, 60

valence, 251, 255

valence bond, 261

valence bond theory, 257, 260, 266

valence shell electron pair repulsion theory, 260–61

van der Waals, Johannes, 289–91

van der Waals's equation, 290

van't Hoff factor, 281

van't Hoff, Jacobus, 175, 247, 278–81, 283, 308, 309–12, 319, 327

Vanini, Lucilio, 89

Venetian method, 65

vibrational motion, 148–50

Viète, François, 87

Vinci, Leonardo da. *See* da Vinci, Leonardo

vinegar, 307

virtuosos, 74

viscosity, 171–72, 309

vital forces, 324

vitamin B_{12}, 331

voltaic battery, 210

von Helmholtz, Hermann. *See* Helmholtz, Hermann von

von Neumann, John. *See* Neumann, John von

Waage, Peter, 178, 276

Waals, Johannes van der. *See* van der Waals, Johannes

Warberg, Emil, 175

Waterston, John, 157, 161–64, 165, 170

Watson, James, 334–35

Watt, James, 133, 148

wave function, 257, 264

wave–particle duality, 359, 360, 364

waves, 346, 374

Weber, Ernst, 340

Weber–Fechner law, 338

Weber's law, 338

weird quantum effects, 355

Wells, H. G., 201, 225

Wieman, Carl, 351

Wien, Wilhelm, 220, 230

Wigner, Eugene, 73

Wilhelmy, Ludwig, 308–309, 310

wine, 309

witch, 75

witchcraft, 58

witchhunts, 75

Wöhler, Friedrich, 144

Woodward, Robert, 268–69

World Set Free, The (Wells), 225

World War I, 225, 240, 252, 301, 303

World War II 291, 329, 334, 346

worms, 104

wortrechnung, 87

Wrinch, Dorothy, 333

xenon, 251

x-ray crystallography, 329–30

x-ray diffraction, 328

x-rays, 221

Yale Univesity, 181

Young, Thomas, 207, 242

zebra's stripes, 346

Zeeman, Pieter, 238, 243, 245, 255

Zeh, Dieter, 365

Zeitschrift für Physikalische Chemie, 281

zero, 47

Zewail, Ahmed, 316

Zhabotinsky, Anatol, 344–45, 346

zinc chloride, 277

zinc hydroxide, 277

Zoroastrians, 46